Strategy in Asia

Strategy in Asia

THE PAST, PRESENT, AND FUTURE OF REGIONAL SECURITY

Edited by Thomas G. Mahnken and Dan Blumenthal

Stanford Security Studies
An Imprint of Stanford University Press
Stanford, California

Stanford University Press
Stanford, California

Printed in the United States of America on acid-free, archival-quality paper

Library of Congress Cataloging-in-Publication Data

Strategy in Asia : the past, present, and future of regional security / edited by Thomas G. Mahnken and Dan Blumenthal.
pages cm
Includes bibliographical references and index.
ISBN 978-0-8047-9149-6 (cloth : alk. paper) — ISBN 978-0-8047-9274-5 (pbk. : alk. paper)
1. Asia—Strategic aspects. 2. National security—Asia. 3. Security, International—Asia. 4. Strategic culture—Asia. 5. Asia—Military policy. I. Mahnken, Thomas G., editor of compilation. II. Blumenthal, Dan, editor of compilation.
UA830.S84 2014
355′.03355—dc23

2014010113

ISBN 978-0-8047-9282-0 (electronic)

Typeset by Newgen in 10/14 Minion

CONTENTS

MAPS

PREFACE

THIS VOLUME GREW out of the recognition of two important facts. The first involves the increasing strategic weight of the Asia-Pacific region, a trend that warrants greater attention from scholars and policy makers alike. The second is focused on the growing importance of the strategic realties of Asia, which require viewing the entire continent under the multiple lenses of geography, culture, history, politics, and economics. Many of the concepts introduced by the advent of the field of strategic studies, including *deterrence* and *arms race*, are similarly quite fruitful.

An understanding of strategic studies can heighten our comprehension of Asia much as a deeper understanding of the region can enhance our knowledge of strategy. Fundamentally, the field of strategic studies is concerned with the relationship between politics and military force. On a basic level, however, strategy is universal: it can be applied in any state or at any time because human nature remains essentially unchanged. Yet strategy is also contextual: strategy is formulated and implemented in specific geographic, cultural, and social environments.

As an academic enterprise, the field of strategic studies arose in the wake of World War II and during the Cold War. The geopolitics of that period directly and indirectly influenced thinking on deterrence, arms races, and military balances. Although students of strategic studies consult Asian classics by Sun Tzu, among others, the field largely springs from Western thinkers, including Thucydides, Tacitus, Clausewitz, Jomini, Mahan, Corbett, Schelling, and Brodie. This volume looks at whether, and to what extent, strategic studies remain valid for Asia of the twenty-first century.

Aaron Friedberg provides an introduction ("Thinking About Strategy in Asia") that clearly places this volume in the context of the historical political-military interaction among nations of the region and their bilateral relationships with the United States.

Geography provides the most fundamental and unchanging context of strategy. Although technology may somewhat eclipse the status of geography, it is unlikely to replace it. Thus, it is appropriate that the initial five chapters explore the Asian landscape as well as the influence of geography on formulating and implementing strategy. Roy Kamphausen (Chapter 1, "Asia as a Warfighting Environment") emphasizes how mountains, deserts, rivers, and other features of Asian geography subdivide that vast area into multiple theaters with distinct characteristics and shape combat on land and sea. It also underscores the relevance of the straits and channels that provide continental Asia with access to open waters. Bruce Elleman (Chapter 2, "The Cyclical Nature of Chinese Sea Power") examines the effects of geography on Chinese strategy during their long history. China has acted principally as a continental power, but it also has exercised maritime power before returning to tend its interests closer to home. Today Beijing is moving into a maritime cycle once again by projecting naval capabilities to unite greater China, expand its economy, and extend its influence across the Western Pacific.

Toshi Yoshihara (Chapter 3, "Chinese Maritime Geography") elaborates on the influence of geography in shaping Chinese thinking on sea power. He stresses that China has come to regard its environment as claustrophobic vis-à-vis islands off the mainland, the so-called first island chain, that constrict its freedom of action. Such geopolitical realities are indispensable to an understanding of Chinese strategic behavior. James Holmes (Chapter 4, "Mahan and the South China Sea") draws on the works of an illustrious American strategist to appreciate the strategic geography of the South China Sea. Finally, Michael Auslin (Chapter 5, "The US Alliance Structure in Asia") traces the enlargement of American security pacts in Asia and outlines the partnerships established by Washington throughout the region. He also examines the threats to those arrangements, including the growth of territorial disputes among Asian powers and the possibility of future disruptions of access to the maritime commons.

A second group of chapters weighs the cultural context of strategy. Colin Gray (Chapter 6, "Strategy and Culture") observes that the concept of culture is problematic as well as necessary to an understanding of strategy. After rehearsing the methodological challenges that distinguish culture as an influence on strategy, he concludes that such barriers should not prevent academics and practitioners from recognizing the importance of culture. Three subsequent chapters provide assessments of how culture has or has not influenced the strategy making of the major powers in the region—namely, China, Japan, and India. These cultural insights are contributed by Andrew Wilson (Chapter 7, "The Chinese Way of War"), S. C. M. Paine (Chapter 8, "The Japanese Way of War"), and Timothy Hoyt (Chapter 9, "The Indian Way of War").

The next two chapters explore the intersection of strategy and economics. For the first time in history, Asia has become the focal point of the global arms market. Over the last two decades, Asian militaries have greatly improved their capabilities. In his contribution on these advances, Richard Bitzinger (Chapter 10, "Military Modernization in Asia") chronicles defense acquisition within the region. Bradford Lee (Chapter 11, "The Economic Context of Strategic Competition") next calculates the link between productivity and competitiveness.

The last chapters investigate the application of the principal concepts of strategic studies to Asia. Possibly no concept was more pervasive during the Cold War than deterrence. According to Michael Chase (Chapter 12, "Nuclear Deterrence in Northeast Asia"), China and North Korea consider deterrence in a way different from how the United States and Soviet Union did during that era. Thomas Mahnken (Chapter 13, "Arms Races and Long-Term Competition") examines advances in US and Chinese defense capabilities and concludes that neither protagonist is strategically autistic. At the same time, their actions fall short of the classic notion of an action-reaction arms race, and their competitors' upgrades in weapons systems are not the exclusive impetus for their modernization efforts.

Then Michael Evans (Chapter 14, "Irregular Warfare in Asia") assesses counterinsurgency theories against their practice in Asia. He claims that strategic studies have addressed irregular warfare in ways that are fraught with undertheorizing and a lack of historical perspective. Furthermore, although much can be learned from the character of the diverse irregular conflicts that have plagued Asia, most of the analytical methods employed in countering insurgency in the region have been grounded in Western strategic studies.

The conclusion to the volume, by Thomas Mahnken, Dan Blumenthal, and Michael Mazza ("Toward a Research Agenda"), looks ahead and identifies some of the most promising avenues for pursuing study and research on strategy in the Asia-Pacific region.

Thomas G. Mahnken
August 2014

ACKNOWLEDGMENTS

THIS VOLUME GREW out of a conference, Asian Strategic Studies, cosponsored by the US Naval War College's Jerome E. Levy Chair of Economic Geography and National Security and the American Enterprise Institute, which was held at the US Naval War College in Newport, Rhode Island, in August 2011. The editors thank all conference participants, including the contributors to this volume. We also acknowledge the stellar assistance of the Naval War College Special Events Department, including Karen Sellers, Shirley Fernandes, and Al Lawton, and of Cathy Hubert of the college's Strategy and Policy Department. We are also grateful to the Naval War College Foundation for its generous support.

We recognize the assistance of Robert A. Silano, who edited the manuscript for publication, and of Ian Cool, who created the maps that enliven the text. We also thank the leadership of the Naval War College, Rear Admiral John Christenson and Ambassador Mary Ann Peters, for their encouragement.

Finally, we thank Geoffrey Burn and James Holt of Stanford University Press for their assistance in bringing this project to print.

ACRONYMS

AAM	air-to-air missile
AIM	airborne interceptor missile
AMRAAM	advanced medium-range air-to-air missile
ASBM	antiship ballistic missile
ASCM	antiship cruise missile
CASC	China Aerospace Science and Technology Corporation
CASIC	China Aerospace Science and Industry Corporation
COIN	counterinsurgency
CT	counterterrorism
DDH	helicopter destroyer
DIA	Defense Intelligence Agency
DPRK	Democratic People's Republic of Korea
GDP	gross domestic product
HMAS	Her Majesty's Australian ship
ICBM	intercontinental ballistic missile
INS	Indian Naval Ship
IRBM	intermediate-range ballistic missile
ISAF	International Security Assistance Force
ISI	Inter-Services Intelligence

IT	information technology
JDAM	Joint Direct Attack Munition
JSF	Joint Strike Fighter
LACM	land-attack cruise missile
LCAC	air-cushioned landing craft
LHD	landing helicopter dock
MRBM	medium-range ballistic missile
PLA	People's Liberation Army
PLAAF	People's Liberation Army Air Force
PLAN	People's Liberation Army Navy
PLANAF	People's Liberation Army Navy Air Force
PRC	People's Republic of China
ROC	Republic of China
RMA	revolution in military affairs
ROK	Republic of Korea
SAM	surface-to-air missile
SMD	Sea-based Midcourse Defense
SLBM	submarine-launched ballistic missile
SSBN	ballistic-missile submarine
SSN	nuclear-powered attack submarine
STOVL	short take-off/vertical landing combat aircraft
SWORD	Small Wars Operations Research Directorate
TFP	total factor productivity
UAV	unmanned aerial vehicle

Strategy in Asia

INTRODUCTION

Thinking About Strategy in Asia

Aaron L. Friedberg

DESPITE THE MARKED DIVERSITY of their topics and perspectives, the chapters in this book form part of a coherent and distinctive intellectual project: to shed light on the past, the present, and above all, the possible future political and military interaction among nations of the Asia-Pacific region and between those nations and the United States. The reason for this focus is apparent. As Henry Kissinger and others have observed, Asia is emerging as the center of gravity in the international system.[1] The rapid economic growth that began with Japan during the 1960s spread to South Korea, Taiwan, and Singapore in the 1970s; China in the 1980s; and India in the 1990s. As has become indisputable, throughout history, prosperity brings power in its train.

Today, Asian nations account for an increasing share of global military resources and overall economic output. Even though defense budgets and force levels have declined in Europe and North America, Asia's have expanded.[2] The region is home to five nuclear-armed militaries (China, India, Pakistan, North Korea, and Russia), and their number could increase. Meanwhile, on the conventional side of the weapons ledger, Asian nations have been investing in advanced combat aircraft, unmanned aerial vehicles, submarines, and surface vessels and progressively expanding arsenals of both long-range ballistic and cruise missiles.[3]

Compared to Europe, Asia has weak international organizations and means of resolving disputes. Moreover, it contains different types of states—from liberal democracies to authoritarian regimes of various stripes and repressive totalitarian dictatorships—with myriad outstanding differences over borders and maritime claims. Asia is also a region in which the domestic politics of many significant players are characterized by strident forms of nationalism. For these reasons, Asia is one region of the world where conflicts among major powers remain plausible and may even be probable.[4] It is also a region where the United States has substantial economic interests, strong alliance commitments, quasi-alliance relationships, and a continuing interest in preserving freedom of navigation across the Western Pacific.

Although it may be obvious *why* students of strategy should care about Asia, questions on *what* to study and *how* to go about it can be somewhat more complex. This book offers answers to those questions; taken as a whole, it provides three elements of a comprehensive program for studying and conducting research on Asia-Pacific strategic issues.

THE FUTURE SECURITY ENVIRONMENT

The future security environment in which the nations of Asia will have to interact includes persistent features of the physical environment as well as material trends and processes that will affect the distribution of power among them but over which no one of them can exert full control.

Geography

Contrary to what Thomas Friedman has maintained, the world is not flat. Geography still matters, certainly in military affairs, and that is nowhere more evident than in Asia. Compared with Europe, the Middle East, and other areas of intense geopolitical interaction, *strategic Asia* is very large; distances within the region are huge, and one key player is more than six thousand miles away.[5] Save for China and Russia, and partly for China and India (which are separated by the Himalayas), the major powers are not physically contiguous. Nations that wish to deter, coerce, or attack enemies must generally be prepared to project power across great expanses of water and airspace, which until recently few were actually capable of doing. Moreover, this is a region in which suitably equipped major powers may fight what Chinese strategists have called *noncontact* wars, engaging one another on the sea and in the air—and perhaps even in space and cyberspace—without ever coming into contact on the land.

With regard to material processes and trends, three stand out as particularly relevant to strategists: demographics, economics, and technology.

Demographics

In terms of its rapid economic development, expanding military establishments, and nationalist identity politics, Asia today resembles early twentieth-century Europe in certain worrisome respects. In one important way, however, it is markedly different. Instead of experiencing rapid population growth and restive youth bulges, many Asian nations, including China, Japan, and South Korea but with the notable exception of India, face aging populations. Others, notably Japan and Russia, will shrink in absolute terms over coming decades. The implications of these demographic trends for economic growth, social cohesion, military policy, and international behavior more generally are unclear, but they could be profound and warrant further study.

Economics

Despite their remarkable performance in recent decades, there is considerable uncertainty about the future trajectories of major Asian economies. India's ability to achieve and maintain annual growth rates closer to 10 percent than 5 percent will go a long way toward determining whether it can achieve its potential to become a true great power. For China meanwhile, the question is when and how rapidly its economic engine will slow. Not even the most optimistic denizens of China's state planning apparatus think that the near-double-digit rates of the last three decades can be sustained indefinitely. What remains to be seen is whether growth slows gradually and gracefully or plummets, perhaps as the result of a crisis involving years of politically motivated overinvestment in real estate and infrastructure.

The ability of China to transition to a more balanced economic development model with greater emphasis on consumption, as opposed to more investment and exports, has significant implications for its national power as well as the welfare of its people. Steady, rapid economic growth has enabled China to expand military budgets without greatly increasing the share of its gross national product devoted to defense. Slower, more erratic progress would mean tougher trade-offs between guns and butter and the likelihood of budget battles among the military services.

Patterns of trade, investment, and infrastructure development within Asia also will have important strategic and economic implications. One possibility is a region in which every road (and pipeline) leads to Beijing, the renminbi is the preferred medium of exchange, and the field for the flow of both goods and capital is tilted in favor of Asian actors and against external competitors. This would clearly be different from a world in which Asia is integrated in a global economy that operates according to liberal trading principles.

Access to and control over natural resources also will drive strategic interaction. The recent intensification of disputes in the South and East China Seas is the most obvious manifestation of this tendency. Economic development and rising living standards have increased demand for food and water as well as energy and minerals. The prospect of scarcity, or even worse, the deliberate denial of resources by hostile competitors, has already become a factor shaping the calculations of planners and decision makers across the region.

Technology

The development and diffusion of strategically relevant technologies will substantially affect the distribution of military power. Nuclear proliferation is the most obvious manifestation of this large and multifaceted process. Although its implications have not become fully apparent, that North Korea has established itself irrevocably as a nuclear weapons state is beginning to register in the minds of the people within the region. The likelihood of South Korea, Japan, and perhaps other nations following suit has always existed in theory, but today it is being considered more openly and taken more seriously than at any time in the past.

Whatever happens in the nuclear domain, more states are obviously determined to acquire the capabilities to project conventional military power beyond their borders. This trend, in turn, fuels interest in antiaccess and area-denial capabilities similar to those that China has developed to counter the preponderance of US military forces. Low-cost drones and cruise missiles launched from land, sea, subsurface, and aerial platforms will threaten naval vessels or commercial ships operating dozens or even hundreds of miles from China's coasts. The proliferation of antiship ballistic missiles could extend defenses even further and affect naval warfare in ways comparable to the advent of carrier aviation in the interwar years. Crowded Asian coastal waters could quickly be transformed into no-go zones in a war, with implications felt around the world. Outside nations that lack a military presence, as do most European powers, could find their interests threatened by developments over which they can exercise little direct control.

State and possibly nonstate actors will have the increasing capacity to launch cyber-attacks. The disruption to South Korean banking and broadcasting in 2013, possibly originating in North Korea, may offer a foretaste of things to come.[6] This form of warfare is likely to be appealing in a region where disputes are deeply rooted, the cost of open conflict remains high, and prosperous, technologically advanced, and powerful nations are the most vulnerable in this dimension.

NATIONAL STRATEGIES

The security environment provides the context in which nations interact. Strategies are plans and programs through which major powers define goals, mobilize resources, and apply those resources to achieve goals, which may vary widely in coherence and integration. At critical moments they will be objects of intense debate and it may be hard to discern whether anything deserving to be called a strategy actually exists. That said, and despite the conceptual, bureaucratic, and domestic obstacles to developing strategies, governments devote considerable energy in trying to behave strategically, and the results of their efforts demand serious analysis.

Contrary to some theories of international relations, the strategy of a given nation cannot be inferred from its relative strength or position in the global order. Ideas, interests, and ideologies as well as external imperatives and material constraints influence strategy. Even the strategies of authoritarian systems are typically by-products of struggles among groups and individuals rather than simply handed down, fully formed, by powerful leaders. To appreciate what nations are doing at any moment and anticipate what they may do, it is necessary to follow elite debates on national strategy. Analysts must examine the logic of the alternatives put forward as well as the coalitions supporting them and the institutional processes and procedures that will determine which alternative, or what amalgam of approaches, emerges victorious.

Australia, China, Japan, South Korea, the United States, and other nations are all engaged in debates of this kind. In Beijing the overarching issue is whether the dictum of Deng Xiaoping that China should "hide [its] capabilities and bide [its] time" has outlived its usefulness and, if so, whether it should be replaced with something more muscular. More concretely, the question facing the new leaders in Beijing is whether to continue the assertive approach to long-running maritime disputes with its neighbors that it began in 2010. In Japan, on the other hand, the question is how best to respond to Chinese forcefulness. The answer, at the moment, seems to involve resistance rather than appeasement. Tokyo has announced plans to increase defense spending and seek tighter strategic cooperation with Washington. It also has taken measures that include relaxing the ban on arms sales to third parties, which are aimed at shoring up the regional balance of power in the face of the current Chinese military buildup.[7]

Australian decision makers and analysts are debating how to manage deepening economic relations with China while preserving their traditional security alliance with the United States.[8] The South Korean military posture and future diplomatic disposition are also in flux. Seoul has already taken steps to loosen American-imposed restrictions on its missile forces, and the issue of an independent nuclear deterrent seems to be back on the table.[9] Even though South Korean elite and public opinion have been growing warmer toward the United States and cooler toward China, relations with Japan remain strained.

Meanwhile, in Washington, debate continues over whether the Obama administration's pivot, or rebalance, toward Asia, an initiative undertaken largely in reaction to Beijing's increasing assertiveness, is stabilizing or provocative. AirSea Battle, which is the integrated warfare doctrine associated with the pivot, has become a source of lively disagreement. Looming above such questions is whether the intensified geopolitical rivalry with China is affordable for the United States given fiscal constraints.

Of the factors at work in Asia, popular nationalism is likely to prove particularly important in shaping national strategies. It would be a mistake to assume, as so much of the political science literature does, that international behavior is produced by rational deliberation and calculation. To the contrary, collective pride and deep-seated animosity, fear, and resentment can play critical roles in shaping national strategy, even when the end results seem obviously counterproductive.

To take one notable example, if Beijing wants to become the dominant power in East Asia once again, it would do well to seek better relations with Tokyo to weaken Japanese ties with Washington. Instead, China has threatened and bullied Japan, driving it further into the arms of the United States. The Chinese Communist Party is promoting anti-Japanese sentiment to bolster its domestic legitimacy, and that complicates efforts to achieve regional hegemony. For their part, Japan and South Korea would be better positioned to cope with the rise of China through closer cooperation. However, passions aroused by the unhappy history of relations between these two countries still make this extremely difficult.

Beyond current national interests and memories of the past are deeper patterns of thought that influence policy makers. China, India, Japan, and other nations have undergone centuries of internal and external conflicts and competition. As a result, they have developed characteristic ways of thinking about politics, diplomacy, and war that differ from those of the West. In their initial interaction with outside powers, Asian societies' obvious material weakness overshadowed their unique strategic cultures. Whatever advantage they might have enjoyed from the subtlety of their statecraft or skill at employing deception in time of war was overwhelmed by the superior strength of their enemies.

The current situation is different, but it is not entirely without precedent. The first wave of scholarly interest in strategic culture in the 1970s coincided with a growing recognition that the United States no longer had a massive edge in military power over the Soviet Union. Albeit belatedly, some American and other Western strategists began to realize their counterparts were not simply laggards who needed to be schooled in the revolutionary effects of nuclear weapons and the virtues of stability. The Soviets had their own approach to warfare, which if put to the test, might have proved superior. In any event, the obvious erosion of previous American advantages made it clear that bolstering deterrence required gaining a better understanding of Soviet thinking. Similarly, the growing strength of China, India, and other Asian nations is kindling a resurgence of interest in their distinctive strategic cultures.

PATTERNS OF STRATEGIC INTERACTION

After examining the board and exploring the plans and goals of key players, students of strategy in Asia will want to stand back and contemplate the evolving pattern of interaction among them. The broadest questions concern the structure of the emerging Asian system and its major axes of antagonism and alignment. Will Asia become really multipolar, with several independent centers of power, including China, India, Japan, Korea, Russia, and perhaps Indonesia, maneuvering with and against one another? Or will the regional system become increasingly bipolar, with a line drawn between China and the United States and like-minded powers, including allies such as Japan and quasi allies like India? Or is Asia—at least East Asia—moving toward a hierarchical order, with China at the center, resembling the premodern tribute system?

These broad structural questions can be broken into two sets of practical, policy-relevant issues. The first involves the management and future of alliances and the possible formation of new, looser groupings of nations that share security concerns even if they do not enter into mutual defense agreements. Established US relationships with Australia, Japan, the Philippines, and South

Korea are all in flux, with a trend toward even closer ties. Nevertheless, the combination of growing concern over Chinese power and the likelihood of persistent downward pressure on US defense budgets means that burden sharing is regaining salience and could become a source of controversy. Efforts by Washington to increase the efficiency of the hub-and-spoke alliance system by promoting greater cooperation among partners also face difficulties, especially in the case of Japan and South Korea. Moreover, the United States is seeking ways to use commercial policy as an instrument of national strategy, proposing free-trade agreements as an alternative to friends and allies being drawn into the orbit of the massive Chinese economy. At the same time, Beijing is attempting to promote alternative regional institutions of its own design that exclude or marginalize Washington.

In addition to transpacific ties, many Asian nations are seeking to forge stronger strategic relationships within their region. The linkages take different forms, including bilateral and multilateral dialogues among participants such as Australia, India, Japan, and Vietnam. Military exercises, intelligence exchanges, and arms sales are also increasing in frequency and volume. Whatever the United States does, Asian nations are seeking ways to work together to shore up their positions in relation to an increasingly powerful China.

Enhanced cooperation in some relationships is being accompanied by intensified military competition in others. Although it has taken time for US officials to acknowledge the obvious, Beijing and Washington have been competing for the better part of two decades. Strategists on both sides regard the other as a potential enemy, which influences deployments, exercises, war plans, research and development, and procurement. While China and the United States are not engaged in a simple action-reaction arms race, each is increasingly focused on the other and their plans are becoming more tightly linked. Each aims to deter the other from taking actions that it opposes and seeks to improve the chances to achieve its military objectives if deterrence fails.

China in particular appears to have adopted a competitive-strategies approach, developing weapons and operational concepts that target US vulnerabilities and will be disproportionately expensive to counter, such as using comparatively inexpensive cruise and ballistic missiles to attack multibillion-dollar aircraft carriers. The irony is that competitive strategies originated in the latter part of the Cold War, and they were intended for use against a

Soviet Union that appeared to be gaining some military advantages. Now the tables have been turned, and it remains to be seen whether, and if so how, the Pentagon can regain those advantages.[10]

• • •

Military competition between China and the United States will not be the only struggle in Asia. China and India observe each other warily across the Himalayas and in the Indian Ocean. China and Japan are not only planning for conflict but maneuvering their forces against one another in the Western Pacific. Additionally, Japan and South Korea are developing capabilities to project power in response to other contingencies, which can possibly be seen as mutually threatening. The nations bordering the South China Sea are enhancing their ability to defend their maritime claims against China, but some have long histories of mutual mistrust. Military interaction in the Asia-Pacific region is complex, multifaceted, and dynamic—and likely to intensify. For better or worse, the study of strategy in Asia will keep scholars and analysts busy for many years to come.

NOTES

1. Henry A. Kissinger, "China: Containment Won't Work," *Washington Post*, June 13, 2005.

2. According to the International Institute for Strategic Studies, the aggregate military spending by nations of the region began to exceed Europe's in 2012. International Institute for Strategic Studies, "Military Balance 2012: Press Statement," March 7, 2012.

3. Regarding the naval dimension, see Geoffrey Till, *Asia's Naval Expansion: An Arms Race in the Making?* (London: International Institute for Strategic Studies, 2012).

4. Aaron L. Friedberg, "Ripe for Rivalry: Prospects for Peace in a Multipolar Asia," *International Security* 18 (Winter 1993–1994): 5–33.

5. *Strategic Asia* refers to the eastern half of the Eurasian landmass and arc of offshore islands in the Western Pacific. This expanse is centered on China and surrounded by four subregions arrayed clockwise around it: Central Asia, Northeast Asia (the Russian Far East, the Korean Peninsula, Japan, and Taiwan), Southeast Asia (continental and maritime domains), and the South Asian subcontinent. See Aaron L. Friedberg, "Introduction," in *Strategic Asia, 2001–02: Power and Purpose*, ed. Richard J. Ellings and Aaron L. Friedberg (Seattle: National Bureau of Asian Research, 2001), 4–7.

6. Choe Sang-hun, "Computer Networks in South Korea Are Paralyzed in Cyberattacks," *New York Times*, March 20, 2013.

7. Mure Dickie, "Japan Relaxes Weapons Export Ban," *Financial Times*, December 27, 2011.

8. For what stimulated the debate, see Hugh White, *The China Choice* (Collingwood, Australia: Black, 2012).

9. Kelsey Davenport, "South Korea Extends Missile Range," *Arms Control Today* 42 (November 2012), http://www.armscontrol.org/act/2012_11/South-Korea-Extends-Missile-Range. In a poll conducted in February 2013, 66 percent of South Koreans favored developing nuclear weapons, while 31 percent were opposed. Jiyoon Kim, Karl Friedhoff, and Chungku Kang, "The Fallout: South Korean Public Opinion Following North Korea's Third Nuclear Test," Asian Institute for Policy Studies, 2013, p. 8.

10. For an introduction, see Thomas G. Mahnken, ed., *Competitive Strategies for the 21st Century: Theory, History, and Practice* (Stanford, CA: Stanford University Press, 2012).

1 ASIA AS A WARFIGHTING ENVIRONMENT

Roy Kamphausen

STRATEGISTS ENDEAVOR to grasp factors that shape potential wartime environments in peacetime to more effectively wage future conflicts. Part of that process involves determining how strategic geography influences warfighting. Additionally, it involves knowing the battlespace to estimate how decisive terrain benefits those who control it, just as analysts study terrain features before wars commence to support military planning.

However, an appreciation of strategic geography alone cannot discover the causes or patterns of war, which does not conform to terrain-based or mechanistic decision models. Nations engage in wars for various reasons and even wage them despite adverse strategic environments. For instance, despite a prevailing view that US strategic interests in Asia are primarily maritime based and related to preserving unfettered freedom of navigation, America conducted four land wars in the region over the past sixty years, including the recent conflicts in Southwest Asia. This suggests that something more than geography influences war and peace. Nonetheless, a process for studying strategic geography can help in understanding a dimension in which military operations are planned and executed.

Certain aspects of Asian strategic geography may be characterized as *decisive*, which describes physical features that offer strategic advantages by establishing conditions for either the success or inhibition of military protagonists. Decisive terrain is not always the most prominent terrain but rather the terrain that provides the military advantage to the side that controls it. In sum, decisive geography is linked to the success of military operations. For that reason, decisive terrain conveys military advantage through means that include physical presence, diplomatic relations, and security alliances.

This chapter supports some preliminary conclusions about Asian strategic geography and its implications for military operations. Moreover, it looks at how man-made features and technological developments can modify or even transform strategic geography. Finally, it provides a net assessment of the problems and benefits that Asian strategic geography, together with decisive terrain, can bring to bear on US and allied militaries.

Asia constitutes the eastern portion of the Eurasian landmass, extending eastward from a north–south line that starts in the Ural Mountains and runs down to the Caspian Sea. It includes Central Asia (Kazakhstan, Kyrgyzstan, Tajikistan, Turkmenistan, and Uzbekistan), South Asia (Afghanistan, Bangladesh, Bhutan, India, Maldives, Nepal, Pakistan, and Sri Lanka), mainland Southeast Asia (Burma, Cambodia, Laos, Malaysia, Thailand, and Vietnam), insular Southeast Asia (Brunei, Indonesia, the Philippines, and Singapore), and North Asia and Northeast Asia (China, Japan, Mongolia, South Korea, North Korea, and Taiwan). It also encompasses partial entities, such as the eastern orientation of Iran and the Russian Far East.

GEOGRAPHIC IMPLICATIONS

The major geographic features of South Asia are mountains that form a belt around both its southern and central subregions and contain all the peaks in the world higher than twenty-three thousand feet. A mountain's importance in military affairs is relative to the surrounding terrain.[1] For millennia the Himalaya and Karakorum Mountains have served as barriers between South Asia and the rest of the continent.

The unique nature of the land geography in South Asia contributed to modern warfare. For instance, the concept of the air bridge first emerged during World War II when Allied transports flew over the Himalayas from India on a route known as the Hump to resupply China. The mountains of South Asia also acted as natural obstacles to the Japanese in southern China. Moreover, Indo-Pakistani wars since Independence have involved large-scale combat operations, but they did not escalate and draw in other nations. Finally, the Sino-Indian War that broke out in 1962 waned before it could turn into a broader conflict, arguably because of the influence of geographic constraints.

Map 1.1. Greater Asia.

The area of this map stretches from the western border of Kazakhstan in Central Asia eastward to the Russian Far East and down to the antipodes in the southern reaches of the Pacific Ocean.

These mountain ranges shape contemporary military operations in two ways. First, channelized terrain and the difficulty of sustaining forces at elevations above sixteen thousand feet impede maneuver warfare. High altitudes and inhospitable terrain in mountainous areas frustrate the rapid advance of armored and mechanized forces in support of conventional military operations. The border war between China and India in 1962 demonstrated the debilitating effects of high elevation on combat operations. Intended by the Chinese as a coercive-punishment campaign to teach India a lesson, the conflict was largely conducted at elevations above sixteen thousand feet and resulted in some six thousand dead or wounded on both sides. Hostilities ended after a month, largely because of the difficulties in sustaining combat operations at those extreme elevations.

Second, the mountain ranges in South Asia also prevent intraregional conflict from becoming broader transregional war even as they can deter such war. The long-standing tensions in Kashmir between India and Pakistan and disputes over Aksai Chin and Arunachal Pradesh between China and India have not resulted in broader conflict, in part because of the limiting effects of the geography. To be sure, the differences are rooted not in the strategic utility of the geography itself but rather in the political borders of the contested areas. The constraints on the escalation of violence that the Himalaya-Karakorum ranges provide notwithstanding, the mountain regions exercise little influence on diminishing tensions.

However, the ways geography might limit the horizontal escalation of a conflict increase the potential for catastrophic vertical escalation, which may take two forms. First, long-range missile strikes may become de rigueur for all protagonists. This action-reaction cycle is occurring with deployments by India of BrahMos missiles in Arunachal Pradesh, by China of short-range ballistic missiles in Tibet, and by Pakistan of short- and intermediate-range missiles. Second, and arguably of greater concern, is that horizontally constrained warfare might lower the bar for using nuclear weapons, especially when major territorial gains might be won at low cost.

NATURAL RESOURCES

Their striking height aside, the decisive terrain of the Himalaya-Karakorum Mountains includes eight great rivers: the Indus, which drains the Himalayas; the Ganges and Brahmaputra, which course through India into Bangladesh; the Ayeyarwady (Irrawaddy), which runs into the Andaman Sea; and the

Salween, Mekong, Yangtze, and Yellow, which flow from the Tibetan plateau. These rivers supply drinking water and power for two billion people.[2] Disputes over water rights among upstream-downstream or other nations with shared boundaries have routinely led to violations of agreements. However, conflict has been prevented by negotiated agreements, such as the India-Pakistan treaty on the Indus and India-Bangladesh negotiations on the Ganges-Brahmaputra river basins or multilateral forums such as the Mekong River Commission, although the commission has been criticized for failing to counter Chinese plans to dam the upper Mekong for hydropower projects.[3]

Controlling rivers has limited military utility. On the tactical level, rivers present obstacles to ground forces, but once forces are across a river, the obstacle becomes much less significant, because a river rarely has an intrinsic value. At a high operational to strategic level, rivers might serve as means of transport and resupply, but this military value is contingent on river depth, currents, riverine obstacles, and so forth. Thus, conflict over rivers for the military value they might represent is a highly dubious concept.

Additionally, modern history offers no examples of wars breaking out between nations over water usage rights.[4] It remains to be seen whether avoiding wars over water resources will continue in the future. Diminished water tables, salinization of freshwater sources from rising sea levels, explosive population increases in areas affected by water shortages, demand by the agricultural sector for more irrigation, and growing need for power generation together might put freshwater supplies in the same position in the twenty-first century that oil enjoyed in the last century. In that case, the circum-Himalaya rivers and the glaciers that feed them could well become decisive terrain in the context of a future Asian war.

The major geographic features of northern and western Asia are its deserts and plateaus. The almost contiguous Taklimakan and Gobi Deserts are the world's third-largest desert expanse. The adjacent Tibetan plateau is the largest and the highest in the world. Together they create a huge sand box conducive to large-scale military operations. If the mountains of South Asia are distinctive for the constraints they impose on warfighting, the deserts and plateaus of northern and western Asia present opposite conditions. Historically, deserts have served as sites of legendary battles, but they do not create inherent military advantages for those who control them. Indeed, their strategic value is best understood as providing the venue for armored warfare that involves closing with and then destroying enemy formations.

For much of the era of Sino-Soviet strategic rivalry, the Gobi Desert posed risks for China. The Maoist strategy of luring in deep enemy forces that might advance from the eastern coast was unsuited to the Chinese border with the Soviet Union, which offered high-speed approaches for armored forces. As a result, the People's Liberation Army was deployed in large defensive formations in the northern (Shenyang and Beijing) and western (Lanzhou) military regions and oriented on the Russo-Mongolian frontier. Beijing shifted attention eastward only after the collapse of the Soviet Union and introduction of border confidence-building measures with Moscow through the Shanghai Cooperation Organization accords. With the Chinese economic center of gravity in thriving cities on the coast, China's strategic orientation shifted to focus on likely challenges to maritime approaches.[5] But just as control of the Himalayas and Tibetan plateau is vital because of rivers, the Gobi Desert could become the focus of strategic interest for minerals that make mining central to the economic development of Mongolia.[6]

DECISIVE GEOGRAPHY

The Himalaya-Karakoram mountain ranges between China and South and Central Asia are the world's tallest mountains, the Yellow and Yangtze Rivers are two of the world's longest rivers, the Gobi Desert is the third-largest desert, and the Tibetan plateau is the largest and the highest (called the *rim of the world*). However, this chapter argues that the most decisive feature of Asian strategic geography is the island chain on its eastern seaboard and straits that dissect the chain, because they shape the economy and security of the mainland.

Maritime Asia contains islands belonging to continental nations that include the Kuriles, which are occupied by Russia; areas of western Borneo, which form part of Malaysia; and Indonesia, Japan, and the Philippines, which together claim more than twenty-seven thousand islands. The islands run from Kamchatka to the Kuriles; through Japan, the Ryukyus, Taiwan, and then the Philippines; and on to Borneo and Indonesia. They extend northwest along the approach to the Malacca Strait near the Nicobar and Andaman Islands. The chain tracks close to the Asian mainland across the ninety-mile Taiwan Strait and as far as a thousand miles away at its northernmost and western points. Moreover, these islands enclose the Andaman, East China, South China, and Yellow Seas, the Seas of Japan and Okhotsk, and the Gulf of Thailand.

Access from the Pacific Ocean to enclosed or marginal seas and the continental landmass is gained through straits and channels. These straits are of two kinds. The first run perpendicular to the Asian landmass and essentially create paths between islands from the continent to the open sea. One example is the Tsugaru Strait, between Honshu and Hokkaido, which became the international corridor used in 2008 by a Chinese People's Liberation Army Navy task force led by a *Sovremenny*-class destroyer. Another is the narrower Ishigaki Strait, between Ishigaki and Miyako Islands in the Ryukyus, which a submerged *Han*-class nuclear submarine traversed in 2004.[7] A second type of strait, such as the Malacca and Taiwan Straits, runs parallel to the continent and offers access from one marginal sea to another.

The straits that pass through and between islands are decisive because they afford military and commercial advantages. The former provide limited offshore control of the mainland, albeit from international waters, and unfettered lines of communication in transporting personnel and supplies. Moreover, straits crisscrossing the first island chain are significant for maritime trade. A large percentage of global trade passes through the South China Sea, which is accessed by the Malacca Strait and the Lombok and Sunda Straits. Eleven million barrels of oil go through the 1.5-mile-wide Malacca Strait every day. China and Japan receive 40 percent of the oil, which accounts for 80 percent of Chinese oil imports.

No strait, not even the Malacca Strait, represents decisive terrain by itself, because substitute transit routes always exist, and lines of communication are not immutable. Rather, the decisive terrain in the region is the aggregate of the island chain facing the eastern edge of the Asian landmass and straits that provide access to the Western Pacific and that extend laterally between the marginal seas. Asia's decisive terrain comprises its maritime features.

CONTROLLING GEOGRAPHY

Decisive terrain can be controlled through direct and indirect means. The former requires forces to dominate such terrain, weapons systems to preclude adversarial use of geography, and patrols and unilateral or multilateral exercises to enhance access. The latter demands relationships with nations that physically control decisive geography through formal alliances and enhanced security partnerships or more flexible political agreements. In the search for control of decisive maritime geography in the Asia-Pacific region, both of these means are evident.

Every maritime nation in Asia is engaged in military modernization. While these programs vary among nations in scope and intensity depending on resources and priorities, they all include decreasing land forces, procuring state-of-the-art combat aircraft, and building up naval capabilities, all of which are facilitated by greater outlays on defense, arms imports, and innovative network-centric information technologies to enhance regional military power.[8]

This pace of regional military modernization notwithstanding, only two nations possess the overall capabilities to control Asia's decisive maritime geography: the nonresident but potent United States and the resident and resurgent China. Although Japan has the military capabilities to control maritime geography, it is restricted by its constitution and by a lack of political will. China and the United States have markedly different approaches to controlling decisive geography.

The United States controls Asia's decisive geography through its alliance relationships, forward bases, and ongoing military presence. In addition to alliances with Australia, Japan, the Philippines, South Korea, and Thailand, the United States also regards Singapore as a major regional security partner. Moreover, though the United States does not have diplomatic relations with Taiwan, it accords that nation the status of a major non-NATO ally.[9]

The United States maintains bases in Japan, South Korea, and the Western Pacific. Other facilities, though not strictly bases, nonetheless support forward-based and transiting forces. Bases in Japan support naval and air components, while those in South Korea primarily deter aggression from North Korea and strengthen extensive strategic interests that include the first island chain. Among the most important of the other locales are the naval and air facilities in Singapore. In addition, there is Utapao Airfield in Thailand, which provides air bridges for contingencies in and outside the region, and the former US bases in the Philippines.

Beyond bases and access points that facilitate peacetime control of the decisive terrain for the United States, the US Pacific Command conducts effective training and bilateral exercise programs; ships of the Seventh Fleet make hundreds of port calls in the region every year. The combined effect of the US military presence reaffirms commitments to a balance of power that supports freedom of navigation, endeavors to prevent the rise of a regional challenge, and stabilizes conditions for economic development in Asia.

China has adopted a cost-imposing, risk-mitigating strategy that does not pursue peacetime control of decisive Asian maritime geography. Instead, it emphasizes the military cost to other nations, specifically the United States, should they try to seize control of maritime geography in a conflict. This approach consists of two main elements. First, China has embarked on naval modernization to project strength while seeking weaknesses in the surface and undersea defenses of the islands near China. These efforts are intended to leverage geographic proximity to the mainland as China pursues ways to escape from the constraints imposed by US control of Asia's decisive maritime geography. The regional agenda being pursued by the People's Liberation Army is matched by an information campaign to shape reassuring perceptions by island nations about Chinese intentions while casting doubt on the US commitment.

Second, China has developed long-range precision-strike weapons, including antiship cruise missiles and so-called carrier-killer DF-21D ballistic missiles, that can leverage interior lines and deny control of or access to decisive terrain by inflicting severe costs on capital ships, especially carriers, which China has decided are critical for US naval forces.[10]

ALTERING GEOGRAPHY

The strategic geography in Asia can be reshaped by other means, including new strategic dimensions such as space and cyber; by man-made alterations to the terrain; and by the introduction of new technologies. Emerging space capabilities can diminish the strategic advantages of the United States to control decisive Asian maritime geography, and China is developing space-based reconnaissance systems to improve over-the-horizon awareness. However, space at best enables conventional operations and offers limited capabilities to control decisive maritime geography.

Similarly, cyber operations are not constrained by geography. They introduce confusion in enemy command-and-control systems and make attribution difficult and thereby contribute to the fog of war. But emerging analysis suggests that cyber may not be the asymmetrical tool it was once alleged to be. Given the limited utility of cyber operations and their role in conventional warfare, the advantage in a cyber engagement could go to the stronger power. Indeed, as a defense expert recently emphasized, "The cyber instrument of war is most likely to be effective when wielded by the strong against the weak in wars for limited aims."[11]

In addition to the ways that cyber and space may alter Asian strategic geography, man-made features also reshape it. Dams, bridges, and tunnels are good examples of man-made features that affect regional strategic geography. But given the decisive maritime geography of the region, the principal man-made feature is the pipeline. If the eleven million barrels of oil that transit the Malacca Strait daily were to almost double, to twenty million barrels, over the next two decades, as has been posited, maritime geography would only increase in importance.[12] Pipelines can mitigate reliance on domestic oil supplies and reduce the risk of transiting areas where China has limited control.

Several pipelines already exist. The first is the Eastern Siberia–Pacific Ocean oil pipeline, which moves three hundred thousand barrels per day to China and another three hundred thousand to Japan and other markets. The second is the oil and natural gas pipeline that stretches from Burma to Yunnan Province.[13] The third goes from Turkmenistan to the Pearl River Delta in China and became operational in 2011. It is the longest liquefied-natural-gas pipeline in the world.[14]

The many benefits of these pipelines include reduced unit costs for the fuel and strengthened regional bilateral relationships for China. In terms of decisive Asian geography, the pipelines reduce the risk that China faces from having almost 80 percent of imported petroleum products transiting the Malacca Strait. However, pipelines eventually become semifixed fixtures of strategic geography, albeit with entirely different risks and vulnerabilities.

The final development that could affect strategic geography in the region is the emergence of new systems that increase reconnaissance dwell times and strike lethality without risk to their operators: unmanned aerial vehicles (UAVs), or drones, and long-range precision strike. The United States has learned the advantages of longer dwell times and lower risks in campaigns in Afghanistan and Iraq. The Chinese have been gaining knowledge from US combat operations while placing high priorities on the development of drones.[15]

Long-range precision strike also may alter the strategic geography of the region. In particular, the types of long-range strategic-strike weapons being developed by China as deterrents may change decisive geography. For instance, new Chinese antiship ballistic missiles could make the straits less decisive by putting naval reinforcement at risk of attack far away from the straits. As James Holmes and Toshi Yoshihara point out, China could launch precision strike from its heartland, taking advantage of strategic depth and thereby complicating any reinforcing response,[16] including to islands within strike range.

INITIAL NET ASSESSMENT

The United States employs a strategy for control of decisive maritime geography that enhances peacetime operational stability while preparing for conflict. It implements this strategy through its unsurpassed naval power that controls the maritime approaches to Asia's decisive geography. US security alliances and strategic partnerships, bases throughout the region, and both bilateral and multilateral exercises with regional partners organized by US Pacific Command are crucial elements of the strategy. Indeed, irrespective of their relations with China, few nations in the region would avoid peacetime engagement with the US military.

The US alliance with South Korea is centered on the eastern edge of the Eurasian landmass and provides a stronger anchor for maritime-based strategy than offshore balancing. The extent to which major regional powers have struggled over this geography suggests that the US–South Korean alliance furnishes impressive advantages without the transactional costs associated with offshore balancing.

For its part, China is pursuing a strategy of deterrence but with few if any opportunities for peacetime displays of military power. Specifically, Chinese military strategy in mid-2011 favored the development of cruise and ballistic missiles, submarines, and mines that deny enemy operations near the coast in a shooting war but do not guarantee control of either the land or the sea. Such capabilities may represent elements of an antiaccess and area-denial approach, which has preoccupied the US Navy, or else a more modest counterintervention (反干涉, *fan ganshe*) strategy.

Chinese strategy may well reflect an assessment of US wartime capabilities, which has the effect of containment in peacetime. But the Chinese strategy in response has limited operational utility because its key elements cannot be employed in peacetime without inviting dire consequences. For instance, striking an aircraft carrier would escalate a crisis and lead to a destructive response. As a result, this strategy centers almost entirely on deterrence as a bluff.

A net assessment finds that the US approach is most likely to control decisive maritime geography. The unsustainable nature of Chinese strategy during combat, as well as its limited peacetime utility, suggests that it is a transitional approach. Over the coming decade, that strategy is expected to look increasingly like that of the United States', with a greater regional presence, more active exercises, and possibly overseas bases. For instance, the People's Liberation Army Navy is already taking a page from the US playbook by deploying its own hospital ship, the *Peace Ark*.

SPHERES OF INFLUENCE?

An assessment of the strategic geography poses important questions for the United States on issues of war and deterrence in the region. American military posture and operating patterns demonstrate a tenacious adherence to what Robyn Lim (in furthering the tradition of Nicholas Spykman and Alfred Thayer Mahan) suggests is the real US strategic interest in Asia. In *The Geopolitics of East Asia*, Lim argues,

> China now represents the chief challenge to the balance of power in East Asia because it possesses the motive, will and opportunity to seek dominance there. If China were to become dominant in East Asia, that would detract from American security directly by excluding US military power from the area, or by seeking to do so; and indirectly by its effects on Japan.[17]

A survey of Asian strategic geography reveals that large portions of the eastern edge of the Eurasian landmass, particularly south of the Taiwan Strait, are neither decisive in the context of military operations nor accessible to offshore powers. Even Thailand, which is America's oldest partner in Southeast Asia, is faced with domestic reservations about US bases despite its support of a closer alliance. This reality implies the existence of a natural "geography of the peace," in which control of the landmass falls to a continental power (China) and control of the maritime domain resides with an offshore naval power (the United States).[18] Proponents of this approach contend that peace through a natural division of spheres of power and influence between continental and maritime affairs is desirable, particularly considering the alternatives. Yet this approach appears to preemptively cede American prerogatives by limiting where it can operate or exert influence. Moreover, it argues for a rollback of US presence, including the bases on the Korean Peninsula, which are located within the continental (Chinese) sphere of influence. This approach calls for major concessions in the US sphere, which were acquired by the United States and its allies at great cost in blood and treasure, in exchange for fleeting guarantees of peace. It is thus debatable whether US dominance of the maritime decisive terrain within the Asia-Pacific region together with access to the Korean Peninsula can meet US strategic objectives writ large.

• • •

Several conclusions can be reached about strategic geography in the Asia-Pacific region. First, although mountains, deserts, and rivers constitute its distinctive features, the decisive strategic geography of Asia remains maritime in nature: the islands, peninsulas, and straits that control access to the economies of Asia's eastern coast. Military advantage is conferred on the power that assembles the more adept and mutually reinforcing set of strategies to control this decisive terrain in a conflict. The two countries best positioned to contest for control of Asia's decisive geography are the United States (an offshore maritime power) and China (the greatest continental power in the region).

Second, decisive terrain in Asia is largely controlled in peacetime by the United States, which has the preponderance of military power with its allies in the region. A range of peacetime activities, including full-spectrum naval capabilities, diplomatic relations, security arrangements, and a strong regional commitment, are widely appreciated as essential to regional stability and fundamental to the postwar economic transformation of the region. These dimensions also bode well for efforts to deter conflicts and, if deterrence fails, to achieve dominant wartime control of this decisive maritime terrain.

Third, Asia's physical geography can be transcended by new technologies, but the effects are probably not as transformational as originally thought. Long-distance pipelines that avoid maritime chokepoints themselves become part of the fixed physical geography with all of the entailed consequences. Additionally, China's burgeoning space power reveals an increasing Chinese reliance on space, which Beijing once saw as a peculiar and glaring weakness of the United States. Indeed, as China relies more on space it may more clearly see its own vulnerabilities, and this could stabilize military uses of space. Moreover, as the military applications of cyber are increasingly understood, the asymmetrical advantages it provides may actually benefit the stronger of powers in a conflict, contrary to current understandings of its military value.

Fourth, the effect of drones and precision strike in a conflict in which both sides employ them would appear to be diminished; their power seems to be in their unilateral employment. If both sides use drones and precision strike, the value of decisive terrain is not necessarily reduced. Rather, they would seem to provide longer-distance means of controlling the existing terrain.

Finally, the United States' current peacetime advantages in controlling decisive terrain in Asia are unlikely to be eclipsed by China, at least in the foreseeable future. Fundamentally, this may reflect differing sources of their respective power. As a maritime nation, America is better prepared to flourish in the decisive maritime geography of the region. Despite its geographic position, China remains limited in its ability to directly challenge the United States, and absent relatively strong naval and maritime allies of its own, China will not achieve more than conditional denial of Asian strategic geography in a conflict.

NOTES

1. Robert Krulwich, "The 'Highest' Spot on Earth?" *Krulwich Wonders* (NPR science blog), April 7, 2007.

2. This conservative estimate contrasts with the claim that Himalayan waters support half the world's population. See Kenneth Pomeranz, "The Great Himalayan Watershed: Agrarian Crisis, Mega-Dams and the Environment," *New Left Review* 58 (July–August 2009): 5–39.

3. Long P. Pham, "Mekong River: The Lancang-Mekong Initiative," *Mouth to Source*, April 20, 2011.

4. Jerome Delli Priscoli and Aaron T. Wolf, *Managing and Transforming Water Conflicts* (Cambridge: Cambridge University Press, 2009).

5. David Finkelstein, "China's National Military Strategy: An Overview of the 'Military Strategic Guidelines,'" in *Right-Sizing the PLA: Exploring the Contours of China's Military*, ed. Roy Kamphausen and Andrew Scobell (Carlisle, PA: Strategic Studies Institute, US Army War College, 2007), 69–140.

6. Credit Lyonnais Securities Asia, "Return to Riches: Mongolia's New Golden Horde," June 2011, pp. 15–35.

7. Peter Dutton, *Scouting, Signaling and Gatekeeping: Chinese Naval Operations in Japanese Waters and the International Legal Implications*, China Maritime Studies 2 (Newport, RI: Naval War College, 2009), 3–6.

8. Richard A. Bitzinger, "Military Modernization in the Asia-Pacific: Assessing New Capabilities," in *Strategic Asia 2010–2011: Asia's Rising Power and America's Continued Purpose*, ed. Ashley J. Tellis, Andrew Marble, and Travis Tanner (Seattle: National Bureau of Asian Research, 2010), 79–112.

9. This status is defined in section 644 (q) of the Foreign Assistance Act of 1961 (22 USC 2403 (q)), 70 FR 50959 (August 29, 2005).

10. Andrew Erickson and Gabe Collins, "China Deploys World's First Long-Range, Land-Based 'Carrier Killer': DF-21D Antiship Ballistic Missile (ASBM) Reaches 'Initial Operational Capability' (IOC)," *China SignPost* 14 (December 26, 2010): 1–24.

11. Thomas G. Mahnken, "Cyber War and Cyber Warfare," in *America's Cyber Future: Security and Prosperity in the Information Age*, ed. Kristin M. Lord and Travis Sharp (Washington, DC: Center for New American Security, 2011), 63.

12. Mikkal Herberg, "Pipeline Politics in Asia: Energy Nationalism and Energy Markets," in *Pipeline Politics in Asia: The Intersection of Demand, Energy Markets, and Supply Routes* (Seattle: National Bureau of Asian Research, 2010), 2.

13. Ibid., 4–7.

14. Wei Tong, "China's Second East-West Natural Gas Pipeline Operated," *China Radio International*, July 1, 2011, http://english.cri.cn/7146/2011/07/01/2702s645786.htm.

15. William Wan and Peter Finn, "Global Race on to Match US Drone Capabilities," *Washington Post*, June 30, 2011.

16. James R. Holmes and Toshi Yoshihara, "Mao's 'Active Defense' Is Turning Offensive," *Proceedings* 137 (April 2011): 298.

17. Robyn Lim, *The Geopolitics of East Asia: The Search for Equilibrium* (London: Routledge, 2003), 2.

18. Robert S. Ross, "The Geography of the Peace: East Asia in the Twenty-First Century," *International Security* 23 (Spring 1999): 81–118.

2 THE CYCLICAL NATURE OF CHINESE SEA POWER

Bruce Elleman

ONE FAIRLY PREVALENT Western view assumes that the geopolitical differences between European maritime and continental powers are constants—once a continental power, always a continental power. This view may date to the Greeks and the twenty-seven-year war between Athens and Sparta when "the Spartan elephant [was] unable to attack the Athenian whale directly and vice versa."[1] Both France and Germany epitomized this view by repeatedly trying and failing to break out of their continental constraints.[2] Moreover, Alfred Thayer Mahan and Halford John Mackinder focused attention on the crucial role of geography in a way that represented maritime versus continental ways of thinking as if they were historical constants.[3]

In Asia, however, there emerged a perhaps more fluid view of how geography and politics interact.[4] A country may undergo periodic cycles as it shifted from being a continental power to a maritime power and back again.[5] In 1957, Lo Jung-pang, who conducted research on maritime affairs in the Song Dynasty for the celebrated series *Science and Civilisation in China*, edited by Joseph Needham, argued that China experienced three maritime cycles during the Qin-Han, Sui-Tang, and Song-Yuan-Ming Dynasties. Lo concluded that after 1949 China was experiencing a fourth cycle under the Republic of China (ROC) on Taiwan and the People's Republic of China (PRC) on the mainland.[6] Although his thesis might have seemed presumptuous in the late 1950s, looking back more than fifty years later, the work by Lo offers a compelling argument that may explain what appears to be the incredible rise of China as a sea power.[7]

Put in historical perspective, the five-hundred-year Chinese maritime cycles are influenced by both internal and external factors. The most relevant domestic factors included the division of China into separate states, each of which created navies to unify the country. That unification process brought China to a new level as a sea power. In time maritime power ebbed as the country was absorbed by internal affairs. Externally, strong neighbors turned the focus of the Chinese away from the sea and toward its borders in the north, west, and southwest. The periods of sea power ended with barbarian invasions that forced a continental focus. Arguably, Beijing is now on the upward swing of this cycle, as it seeks to develop sea power to unite greater China, expand its domestic economy, and extend its power across the Western Pacific.

FOUR CYCLES OF SEA POWER

One explanation of Chinese history construes rising phases of a cycle known as *Yang* and falling phases known as *Yin*, which are male and female attributes, respectively. For example, dynasties often are presented as cycles, with a state going through stages of birth, growth, old age, and death. This cyclical way of thinking goes hand in hand with popularly held Buddhist views in China of death being followed by reincarnation.

With regard to maritime history, Lo detailed in *China as a Sea Power* how the Chinese had attempted but failed to become a sea power at least three times in the past. Those cycles included the Qin (221–206 BC) and Han (206 BC–AD 220) Dynasties; the Sui (581–618) and Tang (618–907) Dynasties; and the Southern Song (1127–1279), Yuan (1279–1368), and early Ming (1368–1644) periods. The voyages of the royal eunuch Zheng He in the early Ming period marked the zenith of attempts to become a sea power. According to Lo Jung-pang, "Periodic shifts of orientation, from land to sea and vice versa, have shaped the course of China's historical, social and cultural development as well as that of her neighbors. China, with her huge population, her territorial vastness and her geographical location occupies a position of dominance in East Asia."[8] At the beginning of the twenty-first century, this thesis is particularly important as China appears to be on the upward slope of a fourth cycle of maritime expansion, one that at this point seems likely to continue until China achieves its goal of becoming a major sea power.

According to the cyclical view of sea power, China has experienced many *Yang* periods of pushing out to sea followed by *Yin* periods in which it was forced by a threatened invasion from the north or an internal problem to pull back. For example, the Qin and Han emperors depended on naval power to not only consolidate their control but also expand abroad. After Emperor Wu of the Han Dynasty consolidated control over the northwest, he assembled a navy to invade southward and destroy the breakaway states of Nan-Yue and Min-Yue. The Han met with more trouble quelling uprisings in Annam (Vietnam) and Korea. But its fleet defeated Annam in AD 42 in "what may have been China's first naval battle against a foreign navy."[9] As Joseph Needham discovered, in the first century the greatest success of the Han "in nautical technology was the cardinal invention of the axial rudder."[10] After a period of gradual decline, during which China was beset by internal problems and threatened by aggressive neighbors, the Han Empire collapsed, foreigners poured in, and China was divided yet again into three competing kingdoms: the Wei in the north, Shu to the west, and Wu in the southeast.

Approximately five centuries later this maritime cycle was repeated. The northern state of Sui invaded the south during the sixth century primarily using naval power. In 589, for example, Emperor Wen used high seas and riverine fleets to destroy the southern Chen Dynasty. This flotilla included huge ships, each with eight hundred fighting men and "superstructure[s] of five decks towering to a height of more than a hundred feet."[11] In a final naval battle that took place on the Yangzi River, not far from modern-day Nanjing, the Sui navy was victorious. Later, the Tang Dynasty used sea power to rule a reunited China, employing naval forces in campaigns against Annam, Champa (now southern Vietnam), Korea, and the Liuqiu Islands. By the middle of the Tang Dynasty, however, China became weak and faced the threat of strong neighbors to the north and west. Eventually, the Tang Empire fell, foreign invaders poured in, and China was divided.

During the third maritime cycle, beginning in the twelfth and thirteenth centuries, it was the coastal states of a divided China that took the lead in building naval and maritime capabilities, which were adopted by the Song Dynasty. Under the Yuan Dynasty, which reunited China by force largely by the adoption of the Song navy, maritime campaigns were launched against Annam, Champa, Java, Korea, and the Liuqiu Islands. During the Ming Dynasty, this sea power peaked when expeditions of Zheng He sailed as far as the Indian Ocean and East Africa. But the naval power of the Chinese was weakened again in the middle of the Ming Dynasty as they became preoccupied by internal affairs and a resurgence of northern nomads.

These three maritime cycles had much in common. The *Yang* upswing by coastal states occurred while China was divided, the height reached as the country was unified, and the *Yin* decline when military power weakened, the people were absorbed by internal affairs, and foreign policy was directed away from the sea toward the north and the west. The transition to continental power and back again lasted five centuries and corresponded to economic cycles.[12] The most dynamic cycle—one many Chinese invoke when examining the current turn of the country to the sea—took place under the Song-Yuan-Ming Dynasties.

SONG-YUAN-MING SEA POWER

Of the maritime cycles, the most recent cycle, during the Song, Yuan, and early Ming Dynasties, was the largest and most impressive. China emerged from civil war and the chaos of the twelfth, thirteenth, and fourteenth centuries as a great sea power embarked on maritime enterprises and significant overseas expansion. It was arguably more of a sea power than a land power, with the Ming navy extending its political influence from Japan to Ceylon and from the Yellow Sea to the Java Sea. Chinese merchants dominated commerce even as Chinese colonists spread southward.

The *Yang* origins of the third maritime cycle are found in the Song Dynasty. Naval forces established by the Southern Song (1127–1279) to resist the Mongol invasion were particularly impressive, reaching a total of twenty-two squadrons and fifty-two thousand men in 1237.[13] Although it is known that the Mongols adopted the Song navy to attack Japan and Vietnam, the Mongols do not easily fit into the cyclical nature of Chinese sea power. "Judged either from the extent of the territories they seized or by the style of life they adopted, the Mongols were initially outside and beyond the age-old rhythm of *Yin* and *Yang* and the geopolitical conditions of the traditional East Asian territory." However, they did succeed in leaving Yuan China with a "territorial unification of even greater extent than those achieved by Han and T'ang."[14] When it suits its purpose, Beijing even cites the Mongol conquests as evidence of its territorial claims; in May 2012, the Foreign Ministry made extensive claims to the South China Sea on the basis of a 1279 survey commissioned by Kublai Khan, the normally much-reviled Mongol conqueror of China.[15]

While not as expansive as the Yuan Dynasty geographically, the naval history of the Ming is usually portrayed as more peaceful. The exploits of Zheng He, in particular, attracted special attention.[16] During a period of thirty years,

Zheng He led seven voyages that demonstrated Chinese naval might and also established political and economic power over maritime Asia. His warships visited states from the Sea of Japan to the east coast of Africa.

Although many Chinese officials and scholars attempt to portray the journeys of Zheng He as friendly, a more recent interpretation concludes that his "armada could land a powerful army at the seaport of any small state the Chinese wished to intimidate."[17] A chief aim of the journeys was to impose Chinese authority. Among his other captives, Zheng He brought back the ruler of a small kingdom near Colombo, Ceylon, and rulers from the Battaks in northern Sumatra. The Chinese also virtually created the kingdom of Malacca, which it made the base of their naval activities in the South China Sea and the Indian Ocean. Normally independent states in Southeast Asia no longer dared to challenge the Chinese within the region.

A major Chinese goal was forcing the states of Southeast Asia to recognize the Ming court. A number of states in the East Indies and southern India derived their position from the emperor, including Palembang on Sumatra, which was ruled by a Chinese governor. China also appointed the governor of Luzon. At one point, according to the official history, forty-two states paid tribute to the Ming.[18] In addition, at least four kings traveled to China in deference to the imperial court. What is more, Japan, Korea, and the Zhongshan kingdom on the Liuqiu Islands off Taiwan accepted Chinese suzerainty. Meanwhile, Annam was invaded and ruled as a province.

The Ming fleet safeguarded sea-lanes in the Malay Archipelago and southern India against piracy. The destruction of an enemy fleet at Palembang by the Cantonese pirate Chen Zuyi was one example. The pirate leader was captured and then returned to China, where he was executed. In payment for help in defeating Chen, another local man who came from Canton, Zhi Jinqing, was made the governor-general of the Palembang territory.[19] This action is proof that Zheng He not only was concerned with trade but was also in charge of military forces. Zheng He also warned Majapahit (part of modern-day Indonesia) and Siam (Thailand) against interfering with merchants and envoys who passed through their waters. Furthermore, on more than one occasion the Ming navy was sent to intervene in foreign disputes by Emperor Yongle: the navy responded in 1403 to an appeal from Jaya Sinhavarman, the king of Champa, whose capital was besieged by Annamese forces. At the appearance of nine Chinese warships, the Annamite fleet withdrew without a fight.

This thirty-year-year sea power period under the early Ming included collecting tribute, controlling piracy, and exerting hegemony over the region. This short period arguably was the peak of Chinese sea power, because with a strong navy China achieved political hegemony over most of maritime Asia. Throughout this period the Ming fleet enjoyed unchallenged dominance over the Yellow, East China, and South China Seas; established a naval base at Malacca; and extended its power to the Indian Ocean and East Africa. China arguably occupied a position of political supremacy over the maritime states of East Asia, with tacit control of sea-lanes and seaborne commerce in Northeast Asia and Chinese emigrants going abroad to establish colonies in the tropical lands of South and Southeast Asia and beyond.

China's third maritime cycle was also larger and longer than the previous cycles. During the three hundred years spanning the Southern Song, Yuan, and early Ming Dynasties, Chinese sea power ushered in a spectacular and far-reaching advance of the nation. In every sense China could truly be called a sea power. To understand the underlying dynamics of how these cycles worked requires an examination of internal and external factors.

INTERNAL FACTORS AFFECTING SEA POWER

Many important internal factors have contributed to increased Chinese sea power, including the reunification and defense of China, the establishment of an overseas economic empire, and the emigration of Chinese, especially to the southern regions. The coastal states of China turned to sea power to help unify the country. In the early Han and Sui Dynasties, sea power played a key role in the process. The founder of the Song Dynasty, Zhao Kuangyin, who had been a military officer, is reported to have skipped his Confucian lessons "to watch naval and marine exercises" and inspect "nearby dockyards, where warships were built."[20] Currently, Beijing is also building a blue-water navy, purportedly to retake Taiwan by force if necessary.

Chinese maritime ascendency during the Southern Song period is a particularly appropriate example, because the country was divided into separate, competing coastal states. The Song was the first of these states to organize a national, permanent, seagoing navy. Driven out of northern China by the invading Jurchen tribe, the Song court was established on the southeastern coast of China. Both the resources and population of this region had promoted strong maritime traditions, which allowed the Southern Song to develop an ocean-going navy.

In addition to unifying the coastal states under the Song, sea power proved important in the defense against northern invaders, in particular the Mongols. Song naval victories off Shandong and at the mouth of the Yangzi in 1161 vindicated the Song belief in their fleet. By employing their navy to the fullest, the Song Chinese were able to resist the encroachment of the Mongol-led forces until the invaders succeeded in building a navy of their own. With their new maritime forces, mainly composed of Song defectors, the Mongols succeeded in crushing the Song defenses on the Han and Yangzi Rivers, capturing the Song capital, and finally, in the historic battle of Yaishan in 1279, annihilating the Song fleet.

The Southern Song navy enabled most of China to be unified by using sea power to fight off northern encroachment for more than a century. Instead of building their own navy, the Mongols essentially adopted the Song navy during their short rule over China. Once the Mongol rule was overthrown in 1368, after less than a century, the Ming Dynasty came to power, and it too largely adopted the Yuan navy. Voyages by Zheng He were conducted soon afterward. The existence of a strong navy in the early Ming Dynasty meant sea power played a crucial role in reunifying China as well as creating an extensive maritime empire. Today, Beijing is closely following in the footsteps of the Ming in trade policies with Southeast Asia.

Creating an Economic Empire

Foreign maritime trade could potentially provide large revenues to the state. Maritime trade flourished during the Han Dynasty, even including indirect trade routes between China and the Roman Empire. During the Tang Dynasty, there was also a growing sea trade among China, Korea, and Japan. Later the reduction of normal sources of revenue during the Song Dynasty caused by nearly constant war obliged the Chinese to look abroad for supplies. The result was the development of maritime commerce on an unprecedented scale.

During the Yuan Dynasty the Mongols continued their maritime trade and expanded on it. Fleets of warships mounted campaigns against Annam, Champa, Formosa, Japan, and Java in ambitious attempts to create an overseas economic empire. For a time, Chinese paper currency was forced on states as distant as Malabar as an "indirect way to obtain gold, jewels, and other treasures from abroad."[21] Chinese ships dominated regional trade. Later expeditions conducted by the Ming, of which those led by Zheng He are best known, extended political and economic influences of China from Japan to Ceylon and from Korea to Java.

The economic might of China abroad reached new heights under the Ming. An upsurge of maritime commerce was not an anomaly but was largely in keeping with the cyclical trends of historical, cultural, and economic development. Internal economic growth culminated in making China into a major sea power in the early Ming Dynasty, just as current growth rates reported by Beijing appear to be replicating this trend. That development, in turn, led to a sharp increase in Chinese emigrants who settled in overseas communities farther south.

Emigration to Foreign Climes

One Chinese characteristic that resisted expansion at sea was introversion, or what some might call isolationism, which was based on the Taoist teaching of withdrawal and remaining generally passive. Arguably, absolutist governments sponsored such attitudes, because people were made pliable and passive, even while opposing beliefs, such as Christianity, that were seen as disruptive to central authority.[22] Even in modern times, Chinese schoolchildren tend to regard sensitive or introverted peers more favorably.[23] Being satisfied with one's lot in life, the desire to remain at home, and a leaning toward pacifism worked against expansionism.

However, counteracting this tendency were the effects of war, when many Han Chinese, fleeing from oppression at home and lured by opportunities abroad, left China to live in foreign countries. The Mongol invasion, in particular, resulted in the first mass exodus by sea in the history of China as thousands of Song refugees flooded southward. Emigrants went abroad to seek political asylum or simply try to make a living. Many of them founded colonies and over time rose to positions of local, and even regional, prominence. Thus, while the overseas empire that ambitious rulers had tried to build proved to be temporary, Chinese emigrants succeeded in founding a more lasting overseas commercial empire.

Most importantly, the Mongol invasion resulted in an exodus of thousands of soldiers and sailors from China. For example, officers and men of the Song army fled to Annam and Champa, the Malay Peninsula, and even to Sumatra. Both Annam and Champa had so many Song soldiers that they were organized into discrete military units. Meanwhile, the destruction of the Chinese fleet at the 1279 battle of Yaishan led to the flight of many Chinese naval officers and sailors to foreign countries. These men took their maritime skills with them, and many of them served in the navies of local rulers, who were anxious to oppose Mongol oppression.

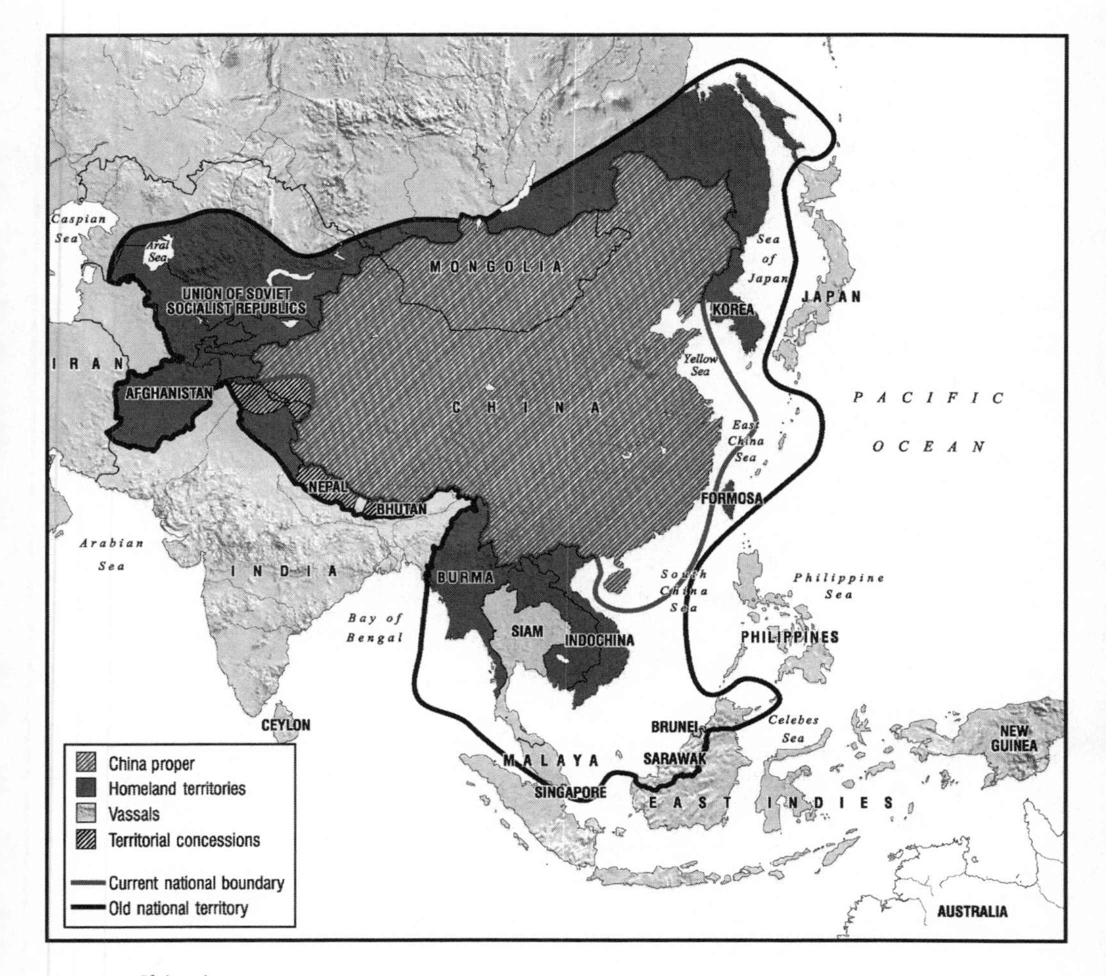

Map 2.1. China in 1927.

This vintage map was adapted from "Zhonghua guochi ditu, zaiban" [Map of China's National Humiliation, Reprint] and displays the extent of China proper under earlier Chinese empires, which included Mongolia. The fifteen homeland territories included Central Asia and Siberia plus Burma, Formosa (Taiwan), French Indochina (Vietnam, Cambodia, and Laos), and Korea. Fifteen vassal states, which sent tribute to China, included Japan, Russia, and Siam (Thailand). Four territorial concessions included two small colonial possessions (Hong Kong and Macao) and two large areas (Kashmir and Nepal/Sikkim). The lines that demarcate the current national boundary and former territorial claims reveal a cycle of periodic moves by China from being a continental power to a maritime power and back again. An original copy of this map, published in Shanghai by the Zhōnghuá Shūjú [中華書局] Book Company in 1927, is held by the Chinese University of Hong Kong.

Throughout the late Song and the Yuan periods, increasing numbers of Chinese emigrants arrived in Southeast Asia to join others who had become established and prosperous. Gradually these refugees grew in strength and on occasion were able to overthrow the native rulers. Large colonies thrived as far away as Sumatra and eastern Java. These Chinese communities remained vibrant well into the Ming period, when Zheng He first visited them.

The interaction of internal forces helped determine the upward cyclical growth of Chinese sea power, including attempts to unify and defend China, economic expansion, and high rates of emigration, especially to Southeast Asia. An extroverted foreign policy fostered the creation of sea power, though at times introversion prevailed and China remained more focused on domestic concerns. The supremacy of sea power during the late Song, Yuan, and early Ming periods was largely the result of the extroverted domestic factors. During the latter half of the Ming Dynasty, however, traditional introversion returned, and treasure fleets were halted. Just as internal factors helped lead China to adopt sea power, external forces turned its attention away from the seas and toward the north, west, and southwest, where threats facilitated by the lengthy and largely unprotected land frontier had so often preoccupied Chinese interests.

EXTERNAL FACTORS AFFECTING SEA POWER

Just as specific internal factors needed to be present to support or inhibit the growth of Chinese sea power, external factors also affected growth. Throughout the history of China, the most important external influences on sea power were strong neighbors along its borders to the north, west, and southwest; invasions and domination by barbarians; and uncontrolled environmental factors such as floods and famines, colder winters, and even desertification.

Threat from the Land

An essential factor in the success or failure of Chinese sea power is the corresponding strength or weakness of land neighbors, primarily to the north and the west and, at times, from the southwest. During periods when the northern and western barbarians were quiet, such as the early Han, Sui, and early Ming Dynasties, the Chinese were interested in furthering sea power. But when border tensions predominated, their focus shifted toward vulnerable land frontiers.

This pattern was evident at the end of the Ming maritime cycle. Renewed concern with the north forced the capital to be relocated from Nanjing in the south to Beijing in the north in 1420. By the time Zheng He returned from his sixth voyage in 1424, the Yongle Emperor had died and his successor, Hongxi, on the advice of continental-minded ministers, ordered the expeditions to the Western Ocean to cease. In particular, Xia Yuanji, the minister of revenue, "criticized the extravagance of Zheng He's maritime treasure fleets and [was] mainly responsible for their abolition."[24] Naval forces abandoned their ships and moved to Nanjing.

In 1425, a new emperor, Xuande, ordered Zheng He to go to sea one last time on a seventh voyage that lasted from 1431 until 1433. This final voyage proved in many ways to be the last gasp in the third great cycle of Chinese sea power. By 1525, Ming imperial edicts had ordered that large ships be destroyed and also outlawed the building of seaworthy junks with more than two masts.[25] From then until the collapse of the Ming Dynasty in the mid-seventeenth century, continental issues trumped maritime ones. To Ming officials, the "sea represented problems, not opportunities, and statecraft stopped, if not at the water's edge, certainly short of the high seas."[26] Chinese efforts to become a sea power were undermined by threats of foreign invasion from the land, just as Beijing today worries that a resurgent Russia might increase the threat from the north.

Foreign Invasion

In addition to the constant threat on China's land borders, actual invasion of China proper by barbarians was the second-most prevalent means of terminating periods of Chinese sea power. For example, the end of the Han Dynasty in AD 220 ushered in a period of dynastic division with the Wei in the north, the Shu in Sichuan, and the Wu in the southeast. For some time, the northern Jin Dynasty reunited China, but it took the Sui Dynasty to resuscitate China entirely. During the Tang Dynasty, in which sea power once again flourished, near constant wars against Turks, Uighurs, and Tibetans drained the dynasty's treasury and contributed to its collapse in 906, which was followed by a chaotic period called the Five Dynasties (Wutai).

The resurgence of sea power in the Southern Song–Yuan–early Ming Dynasties is perhaps the best example of the importance of external threats. The Mongol conquest was an anomaly in the sense that the Yuan Dynasty adopted the Song navy in toto, and attempted to use it to expand the Mongol

Empire even further. In particular, Kublai Khan "aggressively extended his authority overseas, illustrating how quickly the Mongols were able to adapt their military operations to conditions altogether different from those in their steppe homeland."[27] Only naval defeats in wars directed at Japan and Annam halted the spread of Mongol power.

After the fall of the Yuan Dynasty, there was a brief, though strong resurgence of sea power under the Ming. But the land threat reappeared, and Ming emperors became preoccupied with the northern borders. Starting in the mid-fourteenth century, China became more introverted, and treasure fleets were discontinued. As Ray Huang observes, it must have been "absurdly frustrating that not much longer than a hundred years after the voluntary suspension of the seafaring enterprise China had to suffer the presence of Japanese pirates on the eastern coast and see Macao taken over by Portugal."[28] During the next two centuries, the Ming emperors fought against barbarian invasions, which culminated in 1644 with the Manchu conquest of China.

The Environment

Although people, both individually and in groups, were perhaps most important to the cyclical periods of sea versus land power in Chinese history, environmental factors might have also played a crucial role. For example, it appears that a higher than average number of severe winters occurred from 1100 to 1400, approximately corresponding with the invasion of China by the Mongols. Conversely, the maritime cycles appeared to conform to periods of higher than usual temperatures. On other occasions, floods or droughts appear to have similarly important impact on Chinese behavior.

In addition to forcing northern nomads to move farther south, severe winters during these periods meant shorter growing seasons and decreases in agricultural production. Floods and droughts, especially in the southeastern provinces, further aggravated this situation. In time there were huge migrations from the northwest to the southeast, so the coastal provinces became more densely populated. Food production in these areas might have been sufficient under normal conditions, but the added population resulted in greater than usual privation.

For these reasons, environmental factors might have exerted strong pressure for the Chinese to expand outward. One form of expansion was by land, and many Han Chinese continued to push farther southward into what had traditionally been minority (read, non-Han Chinese) lands. But another major form of expansion was by water. Perhaps not surprisingly, it was during this period that China experienced a particularly important shift toward the seas.

The interaction of external forces might have contributed to the *Yin* cycle of Chinese sea power, including the constant threat of invasion from the north, actual invasion and domination of much of China by nomadic peoples, and the unpredictable influence of the environment. It would appear that when the land threat was too great, or when an actual invasion by northern nomads occurred, Chinese sea power was undermined. Conversely, unpredictable environmental factors—including warmer or cooler temperatures than normal, droughts, and floods—could push the Han Chinese to emigrate to the south, which promoted sea power. Arguably, therefore, it was an intricate combination of internal and external factors that contributed to the cyclical nature of Chinese sea power.

• • •

A powerful argument can be made that there is a *Yang-Yin* cycle underlying China's historical periods as a sea power. Although in Europe, certain continental nations—in particular, France and Germany—have experienced relatively short cycles of land power and sea power, in Asia these cycles have appeared to be larger and longer in duration. The recurring shifts of orientation from sea to land and vice versa have shaped the course of China's cultural, social, and economic development, as well as that of many of its Asian neighbors.

On the basis of Lo's arguments for three previous cycles, during the Qin-Han, Sui-Tang, and Song-Yuan-Ming Dynasties, China would appear to be in the midst of the upward swing in the fourth great cycle as a sea power. In a fashion similar to earlier cycles, the People's Liberation Army Navy was organized in 1949 from Nationalist Chinese defectors, with the major objective of reunifying the nation. The outstanding economic growth rates of both Taiwan and the mainland are primarily the result of foreign trade, virtually all of it conducted by sea. Also, largely because of civil war and accompanying chaos surrounding the creation of the People's Republic of China in 1949, millions of people fled to Taiwan, Hong Kong, and Southeast Asia, where approximately ten million Chinese settled.[29] From 1990 to 2010 approximately two and a half million additional Chinese migrants moved into Southeast Asia, bringing with them "capital, technology, cheap Chinese goods and even advanced administration."[30] Xu Liping, a researcher at the Institute of Asia-Pacific Studies in the Chinese Academy of Social Sciences, argues that "having such a large number of overseas Chinese (in the Southeast Asian countries) is a unique advantage of China in boosting people-to-people exchanges."[31] These external factors have all promoted the growth of Chinese sea power.

Conversely, internal factors also could play major roles in this cyclical process. According to the 2010 census, 49.68 percent of the mainland population (665 million people) lives in cities, primarily in eastern regions, which means a rapidly declining population in the central, western, and northeastern areas. In addition, from 2000 to 2010, the percentage of Han Chinese dropped to 91.51 percent, representing a significant increase in minorities.[32] Future ethnic unrest among Mongols, Uighurs, and Tibetans might upset the Chinese push to the sea, just as the threat from Russia or India on the borders or an unexpected environmental change (including rising ocean levels and sharp increase in the danger of massive flooding because of global warming) could again turn the attention of Beijing toward the land.

China would appear to be in the initial uphill stage of this maritime cycle, as the mainland and Taiwan use sea power to compete for dominance. As in earlier maritime cycles, the primary People's Liberation Army Navy goal remains unification. Moreover, support of the overseas PRC economic empire also increasingly concerns Beijing. Although some argue that "China's naval buildup will not pose a challenge to US maritime security," this view may focus on the here and now and not on the future when Chinese sea power may take on a global dimension.[33]

History never repeats itself exactly, but this cyclical view of Chinese sea power brings up thought-provoking questions about previous attempts and subsequent failures to become a true sea power. For example, will China use its resurgent sea power to retake territory by force that it deems lost to foreign interlopers, such as the South China Sea? Or as Lo suggested, will the growth of sea power produce a more peaceful, prosperous, and cooperative nation willing to seek economic integration with its neighbors rather than military competition?[34]

A strong and outward-looking China is once again developing its naval forces and seeking maritime dominance across the region that could have significant implications for global trade, international relations, and military affairs. Its extensive territorial claims in the South China Sea are especially worrisome. But barring unforeseen land-based or environmental threats, the fourth Chinese maritime cycle might be seen one day as its last, relegating the nation to the pantheon of sea powers that have been epitomized by Britain, Japan, and the United States.

NOTES

1. Karl Walling, "War and Leadership: A Critical Analysis of Thucydides' Account of the Athenian Expedition to Sicily," *FPRI Footnotes* 12 (November 2007), http://www.fpri.org/articles/2007/11/war-and-leadership-critical-analysis-thucydides-account-athenian-expedition-sicily.

2. On France, see Geoffrey Symcox, *The Crisis of French Sea Power, 1688–1697* (The Hague: Martinus Nijhoff, 1974); on Germany, see Holger H. Herwig, "The Failure of German Sea Power, 1914–1945: Mahan, Tirpitz, and Raeder Reconsidered," *International History Review* 10 (February 1988): 1–172.

3. See Alfred Thayer Mahan, *The Influence of Sea Power upon History, 1660–1783* (Cambridge: Cambridge University Press, 1890); and Halford John Mackinder, "The Geographical Pivot of History," *Geographical Journal* 23 (1904): 421–437.

4. Edwin O. Reischauer and John K. Fairbank, *East Asia: The Great Tradition* (Boston: Houghton Mifflin, 1960), 114–23.

5. Toynbee focused on economic and military cycles rather than on continental versus sea power. Arnold J. Toynbee, *A Study of History* (New York: Oxford University Press, 1957), 2:298–303.

6. Lo Jung-pang, *China as a Sea Power, 1127–1368: A Preliminary Survey of the Maritime Expansion and Naval Exploits of the Chinese People During the Southern Sung and Yuan Periods*, ed. Bruce A. Elleman (Singapore: National University of Singapore Press, 2012).

7. See James C. Bussert and Bruce A. Elleman, *People's Liberation Army Navy: Combat Systems Technology, 1949–2010* (Annapolis, MD: Naval Institute Press, 2011), 1–17.

8. Lo, *China as a Sea Power*, 342–43.

9. Ibid., 37.

10. Colin A. Ronan, *The Shorter Science and Civilisation in China: An Abridgement of Joseph Needham's Original Text* (London: Cambridge University Press, 1978), 1:39.

11. Arthur F. Wright, *The Sui Dynasty* (New York: Alfred A. Knopf, 1978), 148.

12. Chi Ch'ao-ting, *Key Economic Areas in Chinese History: As Revealed in the Development of Public Works for Water-Control* (London: Allen and Unwin, 1936), 9–10, outlines three periods, each roughly 500 years long: 559 BC (Wu) to AD 42 (later Han), 581 years; AD 226 (Wu) to AD 670 (Tang), 444 years; and AD 907 (Wu–Yue) to AD 1480 (Ming), 543 years.

13. Jacques Gernet, *Daily Life in China on the Eve of the Mongol Invasion, 1250–1276* (Stanford, CA: Stanford University Press, 1962), 72.

14. Gari Ledyard, "Yin and Yang in the China-Manchuria-Korea Triangle," in *China Among Equals: The Middle Kingdom and Its Neighbors, 10th–14th Centuries*, ed. Morris Rossabi (Berkeley: University of California Press, 1983).

15. Jane Perlez, "Beijing Exhibiting New Assertiveness in the South China Sea," *New York Times*, May 31, 2012.

16. Louise Lavathes, *When China Ruled the Seas: The Treasure Fleet of the Dragon Throne, 1405–1433* (New York: Oxford University Press, 1996); Gavin Menzies, *1421: The Year China Discovered America* (New York: Harper Perennial, 2004); Edward L. Dreyer, *Zheng He: China and Oceans in the Early Ming Dynasty, 1405–1433* (London: Longman, 2006).

17. Dreyer, *Zheng He*, 184.

18. Lo, *China as a Sea Power*, 338.

19. Bruce Swanson, *Eighth Voyage of the Dragon: A History of China's Quest for Seapower* (Annapolis, MD: Naval Institute Press, 1982), 39.

20. Ray Huang, *China: A Macro History* (Armonk, NY: M. E. Sharpe, 1989), 107.

21. Lo, *China as a Sea Power*, 317.

22. Immanuel C. Y. Hsu, *The Rise of Modern China* (New York: Oxford University Press, 1995), 388.

23. Xinyin Chen, Kenneth H. Rubin, and Yuerong Sun, "Social Reputation and Peer Relationships in Chinese and Canadian Children: A Cross-cultural Study," *Child Development* 63 (1992): 1336–43.

24. Peter C. Perdue, *China Marches West: The Qing Conquest of Central Eurasia* (Cambridge, MA: Harvard University Press, 2005), 56.

25. Swanson, *Eighth Voyage of the Dragon*, 43.

26. Jonathan D. Spence and John E. Willis, Jr., eds., *From Ming to Ch'ing: Conquest, Region, and Continuity in Seventeenth-Century China* (New Haven, CT: Yale University Press, 1979), 215.

27. Charles O. Hucker, *China's Imperial Past: An Introduction to Chinese History and Culture* (Stanford, CA: Stanford University Press, 1975), 285.

28. Huang, *China: A Macro History*, 156.

29. G. William Skinner, "Overseas Chinese in Southeast Asia," *Annals of the American Academy of Political and Social Science* 321 (January 1959): 136–147.

30. Zhuang Guotu and Wang Wangbo, "Migration and Trade: The Role of Overseas Chinese in Economic Relations Between China and Southeast Asia," *International Journal of China Studies* 1 (January 2010): 174–93.

31. Xu Liping, "China to Boost Ties with ASEAN, Eying Common Development," *Xinhua*, October 9, 2013.

32. "China's Mainland Population Grows to 1.3397 Billion in 2010: Census Data," *English News*, April 28, 2011, http://news.xinhuanet.com/english2010/china/2011-04/28/c_13849795.htm. Interestingly, 2010 data showed the total population was 1.37 billion, but only 1.34 billion were living on the mainland, which means that Beijing is including some 30 million overseas Chinese as citizens, many living in Taiwan, Hong Kong, Singapore, and throughout Southeast Asia, who are considered by local governments to be citizens of those countries.

33. Robert S. Ross, "China's Naval Nationalism: Sources, Prospects, and the US Response," *International Security* 34 (Fall 2009): 75.

34. Lo, *China as a Sea Power.*

3 CHINESE MARITIME GEOGRAPHY

Toshi Yoshihara

CHINA'S SEAWARD TURN has emerged as one of the most anticipated geopolitical events of the early twenty-first century. Thus far, most analysis has been focused more narrowly on naval developments such as carriers, submarines, and antiship ballistic missiles. Although operational capabilities are indispensable to understanding Chinese maritime power, they do not reveal the whole story. According to Alfred Thayer Mahan, sea power relies on immutable natural conditions, including maritime geography.[1]

When the Chinese think about the future of their national maritime endeavor, they envision a prolonged struggle in a complex geographic environment. They visualize their surroundings in claustrophobic terms. To Beijing, the *first island chain*—the long ring of islands lying just off the eastern end of Eurasia—hems in the Chinese mainland. This geographic view is integral to Chinese strategic thinking on sea power.

THE FIRST ISLAND CHAIN

Chinese commentators trace the origins of the island chain concept to US strategic thought during the early years of the Cold War. They consider this archipelagic framework as the most concrete expression of American hostility toward the newly established People's Republic of China. They quote Secretary of State Dean Acheson, who sketched a defense perimeter of the Pacific that ran along the Aleutians, Japan, Okinawa, and the Philippines.[2] These analysts blame other architects of Cold War policies for providing substance to the idea of an island chain as a geographic barrier designed to thwart the expansion of communism from the Asian mainland. Liu Hong, for example, has mentioned an address to a joint session of Congress in 1951 by General Douglas MacArthur in which he argued that control over the islands from the Aleutians to the Marianas would enable the United States to dominate the Pacific from Vladivostok to Singapore with naval and air forces.[3]

Sang Hong refers to testimony by John Foster Dulles before the Senate Foreign Relations Committee that Taiwan embodied an "important link in the so-called 'island chain' that bounds the western rim of the Pacific."[4] Chen Chungen and Jiang Sihai observe that President Dwight Eisenhower warned of Taiwan falling into unfriendly hands, claiming the loss would break "the island chain of the Western Pacific that constitutes, for the United States and other free nations, the geographical backbone of their security structure in that ocean."[5]

To some Chinese observers, the architecture of US military power in the Pacific is directly attributable to the hoary strategy of containment. As Huang Yingxu of the Chinese Academy of Military Science contends, "The US assembled a C-shaped strategic formation" incorporating "the first and second island chains formed in the 1950s." In this view, the United States has transposed its Cold War containment strategy to the post–Cold War era by inscribing a C-shaped encirclement, or arc, on Eurasia. While this strategy "may not be entirely aimed at China," Huang continues, "it surely has the intention to curb and contain China."[6] Bad memories die hard.

Such commentary on US strategic thought during the early Cold War period is certainly based on an invented past, the product of Beijing's chattering class. While it may be true that policy makers in Washington explicitly referred to an island chain in public statements and official documents in the 1950s, they were clearly employing the term as an evocative metaphor to underscore the value of island strongholds in maritime Asia. There is little evidence that this term of art carried strategic or operational weight in policy deliberations in Washington. Indeed, the concept of an island chain regained theoretical salience only after Western analysts detected an emerging Chinese naval debate during the late 1980s.[7] They were simply following the cue of the Chinese in this regard, not the other way around.

The Chinese version of events, however, reveals far more about their worldview and their thinking about US strategy. In retailing this Cold War narrative, Beijing tells a politically correct story about their maritime environment and corresponding strategic choices. According to this account, Chinese naval ambitions are simply a reaction to a highly aggressive and threatening

US military posture in Asia. In short, a selective reading of history has served as a convenient analytical vehicle for framing the current debate on Chinese maritime strategy.

Although it is unclear exactly when the term *first island chain* entered China's lexicon, it most likely gained currency in the 1980s when Admiral Liu Huaqing was commander of the People's Liberation Army Navy. Deng Xiaoping appointed Liu to the post in 1982 to reform the service following the chaos of the Cultural Revolution. In a speech at a 1987 symposium on naval developments, Liu stated, "The first island chain refers to the Aleutian islands, Kurile islands, the Japanese archipelago, the Ryukyu islands, Taiwan island, the Philippine archipelago, and the Greater Sunda islands in the Western Pacific that form an arc-shaped arrangement of the islands akin to a metal chain."[8] He viewed the island chain in expansive terms, stretching across vast bodies of water from the North Pacific to the heart of the South China Sea.

The year before the speech, Admiral Liu issued a National Defense University report that for the first time articulated the basis of a coherent naval strategy. In his masterful analysis Liu asserted that the first island chain delineated the scope of naval operations, comprising "the wide sea areas west of the Japanese archipelago, the Ryukyu Islands, and the Philippines islands."[9] Admiral Liu further declared that the waters enclosed by these islands include China's exclusive economic zones and Chinese-claimed territories in the South China Sea. Thus, to Liu, the island chain not only established the operational area of responsibility for the navy but also represented the sum total of the Chinese economic and territorial prerogatives in the maritime domain. Today *first island chain* is a basic term in the Chinese strategic lexicon. Indeed, a full-text search of the China National Knowledge Infrastructure database of academic journals reveals more than six hundred articles published in the last decade using the term.[10]

ISLANDS EVERYWHERE

In a comprehensive survey of Chinese maritime geography, *Island Chain Surrounding China*, Liu Baoyin and Yang Xiaomei define the first island chain as an *island belt* connecting the Greater Sunda Islands, Japanese archipelago, Philippine archipelago, Ryukyu Islands, and Taiwan—a "crescent-shaped island chain . . . interlocked along [the] nation's coastal areas."[11] They claim, "This geographic conformation whereby an island chain separates a continent from an ocean is the only one of its kind in the world."[12] The islands form a

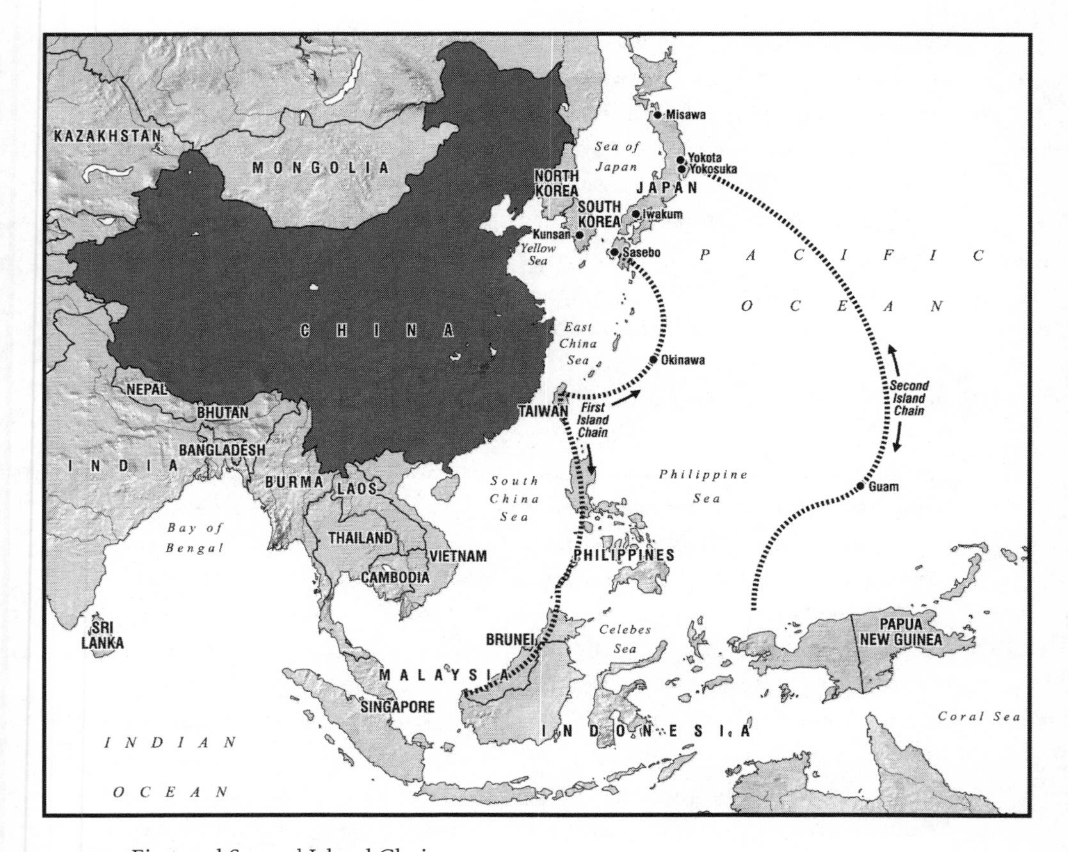

Map 3.1. First and Second Island Chains.
The first island chain has been described as a crescent-shaped grouping of islands that includes the Aleutian Islands, Kurile Islands, Japanese archipelago, Ryukyu Islands, Taiwan, Philippine archipelago, and Greater Sunda Islands within the Malay Archipelago. The second island chain runs from the Japanese archipelago down to the Bonin (Ogasawara) and Marshall Islands.

series of straits and channels through which Chinese mariners pass to reach the world's oceans. Liu and Yang list twenty-two straits and channels, from the Soya Strait in the north to the Palawan Strait in the south, that are critical to China's national security and economic development.[13]

For three naval combat-systems engineers from the Chinese Marine Design and Research Institute, the first island chain animates Beijing's threat assessments. First, given that major straits and channels along it are under the control of other states, Chinese sea trade is vulnerable to blockades at critical chokepoints. Second, Chinese competition for resources above and beneath the continental shelf and other territorial disputes with its neighbors are located within or near the island chain. Third, given the proximity of the mainland to the island chain, many Chinese coastal cities are situated within reach of long-range, precision-strike weapons deployed along the archipelago. They conclude, "Our maritime frontiers lack strategic depth, permitting our nation's economically advanced regions along the coast to directly face enemy threats."[14]

It is not surprising, then, that when geopolitically minded strategists gaze eastward across the seas, many see an island barrier obstructing their nation's entry into the oceanic thoroughfare. In the eyes of some analysts, the first island chain compromises the long coastline of mainland China by restricting nautical activities. Writing in the journal *China Military Science*, Feng Liang and Duan Tingzhi of the People's Liberation Army Naval Command College depict the apparent encirclement in graphic terms, observing that the "islands obstruct China's reach to the sea. . . . The partially sealed-off nature of China's maritime region has clearly brought about negative effects in China's maritime security. . . . Because of the nature of geography, China can be easily blockaded and cut off from the sea, and Chinese coastal defense forces are difficult to concentrate."[15]

Concurring, Major General Peng Guangqian of the Academy of Military Science laments, "Even though our nation is a great littoral power, the sea areas surrounding our nation are either sealed off or semi-sealed off. . . . This has further added strategic pressure from the seas upon China while increasing the difficulty and complexity of China's maritime defense."[16] Interestingly, Senior Colonel Wang Chuanyou likened China's geostrategic position to that of Germany's in two world wars. To Wang, the British Isles, Orkney Islands, and Shetland Islands formed a mini-island chain across the North Sea, blocking German entry to the Atlantic.[17] Although that historical analogy is not precise or particularly satisfying, it nevertheless reflects an intellectual effort to inform thinking on Chinese maritime geography. Together, these strategists' works convey a persistent wish to escape from what they regard as a maritime straitjacket.

This widespread recognition that China faces a competitive and potentially unfavorable maritime environment reflects a larger debate about Beijing's future geostrategic choices. In contrast to simplistic Western depictions of China as a land-bound continental power, Chinese analysts possess a far more nuanced understanding of their nation's geographic orientation.

Given that China borders fourteen nations and faces six maritime states along its eleven-thousand-mile coastline, commentators characterize China as a composite land and sea power. Beijing thus faces the challenge of balancing its strategic demands, which pull its attention in divergent directions.

In the late 1990s Chinese analysts began to search the past to inform the present and future. Shao Yongling and Shi Yanhong, researchers at the Second Artillery Command College and the Institute of International Relations of the People's Liberation Army (PLA), respectively, analyzed the rise and fall of European great powers that share China's geographic predicament. To them, the experiences of France, the Netherlands, Portugal, and Spain represent the most apt analogies to the Chinese situation. Shao and Shi argue that "a hybrid land-sea power" must avoid simultaneous two-front threats on land and at sea at all costs. They also conclude, "As a composite land-sea power, if China seeks to develop at sea, it must ensure continental stability. This is the premise of strategic concentration."[18]

Shen Weilie of the National Defense University draws three lines of geostrategic fronts in northern, southwestern, and eastern peripheries of China to assess the geopolitical environment. Shen contends that landward security challenges to the north and southwest largely abated with the breakup of the Soviet Union, improved relations with India, and friendlier ties with smaller nations. By contrast, the Chinese eastern maritime frontier, which is fraught with flashpoints on the Korean Peninsula, across the Taiwan Strait, and in the South China Sea, is highly unstable, with US hegemony posing an unremitting threat. As Shen concludes, "China's eastern coastal geostrategic front is the most severe situation facing 21st century security and is the main battle line that endangers China's security."[19]

Li Yihu of Beijing University notes, "China's high economic growth regions are located near the riskier areas facing east although they are far from the buffer zone facing west. To a certain extent, China's westward geopolitical situation is more in line with post–Cold War characteristics than the eastward environment. This has created a degree of difficulty or, to put it another way, a degree of risk to China's geostrategic choices between eastern and western directions."[20] Obviously, Chinese scholars are acutely aware of the potentially stark geostrategic trade-offs, as Beijing looks seaward.

Chinese analysts agree that their nation must deal with many omnidirectional geopolitical pressures, but they often differ sharply over geostrategic priorities and appropriate responses. Tang Yongsheng, a professor at the

National Defense University, believes that China faces an insurmountable strategic challenge to its east: American naval dominance. In other words, China is already locked out of maritime Asia. Thus, Beijing must turn west, toward the heartland of Asia, to "greatly alleviate the strategic pressure coming from the east."[21] Liu Zhongmin of Shanghai International Studies University advances a more tempered idea that the constraints inherent in a hybrid land-sea power and weak comprehensive national power compel Beijing to pursue the development of limited sea power.[22]

Kong Xiaohui of Renmin University in Beijing urges a more forward-leaning posture at sea, arguing that developing sea power is a necessary strategic choice. Yet sea power enthusiasts such as Kong acknowledge the need to balance continental and maritime prerogatives. As she explains, a dialectical relationship exists in which "land power safeguards sea power, while sea power is an extension of land power. Without sea power, China's land power cannot be guaranteed; without land power China's sea power loses the foundation for its very existence."[23]

Beijing's choices are not easy or self-evident. It is against this contentious but intellectually stimulating debate that Chinese analysts assess the impact of the first island chain.

TAIWAN: THE CENTRAL LINK

Although the policy community in the United States remains deeply divided over the geostrategic importance of Taiwan, the notion that the island has strategic and military value is relatively uncontroversial on the mainland. Indeed, the intersection of geography and strategy is central to narratives about Taiwan's place in the first island chain. An authoritative PLA document, *The Science of Military Strategy*, captures the essence of this reasoning:

> If Taiwan should be alienated from the mainland, not only [will] our natural maritime defense system . . . lose its depth, opening a sea gateway to the outside forces, but also a large area of water territory and rich reserves of ocean resources [will] fall into the hands of others. What's more, our line of foreign trade and transportation[,] which is vital to China's opening up and economic development[,] will be exposed to the surveillance and threats of separatist and enemy forces.[24]

Other Chinese analyses have elaborated on the geostrategic features of Taiwan as an organic and indispensable component of a maritime frontier that overlaps with the first island chain.

In a study sponsored by the China Institute for International Strategic Studies, Wang Wei depicts Taiwan in precise geometric terms. Along China's 11,000-mile coastline, Hainan Island, the Shandong Peninsula, and Taiwan are the maximum seaward extensions of Chinese territory. The distances from the tip of the Shandong Peninsula to Kaohsiung in Taiwan and to Yulin in Hainan are almost the same, about 750 nautical miles. For Wang, the three protruding points align to form a maritime defense perimeter in the shape of an isosceles triangle with Taiwan at the apex, sitting astride China's north–south line of communications.[25]

In theory, military capabilities in Hainan, Shandong, and Taiwan could mutually support operations in offshore waters and beyond. Without Taiwan, however, as two analysts at the PLA University of Foreign Languages state, "China's maritime defenses would be cut into two pieces while our navy would be forced to operate separately in the two seas, unable to provide mutual support."[26] Moreover, Zhang Shirong of the Central Committee Party School claims that if "Hainan Island loses mutual support from Taiwan Island, the defense of the Spratly Islands [will] erode, making the protection of sea rights in the Spratlys far more difficult."[27]

History in part fuels concerns that Taiwan might impede coordination among China's three fleets. In the 1974 Sino–South Vietnamese naval clash over the Paracel Islands, the PLA Navy (PLAN) dispatched three vessels from the East Sea Fleet to reinforce the South Sea Fleet. At that time, Chinese naval ships typically avoided the Taiwan Strait, preferring the detour around Taiwan's east coast to reach the South China Sea. However, Mao Zedong ordered the warships to transit the strait, despite the concern that Nationalist forces on Taiwan and the offshore islands might ambush the flotilla. Mao's gamble paid off when the frigates arrived safely after an uneventful voyage. Although most fighting had ended, the belated naval show of force deterred the South Vietnamese from further escalation, thereby establishing the Chinese position in the Paracels. This victory and its lessons remain a staple in Chinese historiography.[28]

As a geographic marker, the Taiwan Strait also symbolizes the asymmetrical structure of Chinese seaborne commerce. If Quanzhou, near the midpoint of the strait, is used to divide the mainland coast along a north–south axis, most ports are found in the north. Yet three of the four main international shipping routes leading to Southeast Asia, Oceania, Europe, and the Western Hemisphere head southward.[29] Only vessels en route to the Korean Peninsula and Russian Far East wend their way north. Consequently, the return of Taiwan to China would redress the imbalance between the economic access points and the direction of seaborne trade. In the words of Li Jie, "Possessing Taiwan opens an advantageous waterway to the interior seas of the second island chain while opening a convenient path to the high seas. As such, Taiwan Island serves an important function as the central pivot of the first island chain."[30]

Zhan Huayun notes that the major Japanese islands border the East China and Yellow Seas while Southeast Asia encloses the South China Sea. Accordingly, "Taiwan's ocean facing side on the east is the only direct sea entrance to the Pacific."[31] As Zhu Tingchang of the PLA Institute of International Relations explains the geostrategic relevance of Taiwan, "For China to develop in the Pacific, it must charge out of the first island chain. And the key to charging out of the first island chain is Taiwan. . . . [That] is akin to a lock around the neck of a great dragon."[32] Hyperbole aside, this view of Taiwan as a nautical opening is widely shared.

THE NORTHERN ANCHOR

To Taiwan's north, the lengthy Japanese archipelago occupies the strategic intersection between the maritime interests of rival powers. As Zhang Songfeng of the PLA Institute of International Relations observes, "The maritime lifeline that Japan depends upon for its imports and exports is also the only passageway for China's eastward entry into the Pacific, the US westward entry into East Asia, and Russia's southward movement."[33] To Liu and Yang, the Russo-Japanese War vividly displayed the confluence of Japanese geography and great power struggles in Northeast Asia. Exploiting Japan's advantageous geographic position, the Imperial Japanese Navy kept the Russian Asiatic squadrons divided and confined to Port Arthur and Vladivostok.[34] The inability of Saint Petersburg to concentrate its fleet was a key ingredient to its defeat.

Liu and Yang describe the Japanese islands as an impassable maritime great wall and argue that the proximity of the archipelago to Eurasia means that military power can be projected along and into the continent. They also observe that the "combat radius of advanced fighters launched from bases on the Japanese home islands could reach the interior of East Asia. Warships that sortie from Japanese ports could conduct operations along the East Asian littoral without refueling en route."[35] Not surprisingly, that the Japanese archipelago is home to the combined military power of the United States and

Japan informs Chinese assessments of US forward presence. Feng and Duan also argue that "Japan's current oceanic security strategy relies on an oceanic alliance based on Japan-US sea power cooperation. . . . Moreover, both possess favorable geographic advantages arising from island chain encirclement, a posture that can easily pressure China from the oceanic direction."[36] They see a strategic bloc that possesses the resolve, capability, and geographic position to frustrate Chinese maritime ambitions.

Geopolitically minded commentators pay special attention to the Ryukyu Islands. From an economic perspective, Chinese shipping depends heavily on the Osumi and the Miyako Straits. The vast majority of seaborne traffic connecting Shanghai, Ningbo, and Hong Kong to markets in the United States and Canada passes through Osumi, located just south of Kyushu. Boasting the most direct access to the great-circle route, which cuts down transit distances by more than six hundred miles and transports approximately a quarter of Sino-American trade, Osumi is a preferred gateway even for southern Chinese ports.[37] Commercial shipping of Chinese goods to Oceania and Central and South America often transits the Miyako Strait. The Ryukyu Islands are thus central to Chinese trade through intra- and transpacific commerce.

From a military perspective, there is concern that this crescent-shaped archipelago closes off the Chinese mainland from the Pacific. As Zhang Xiaowen notes, "The surrounding seas of Japan's so-called 'southwestern islands' (referring to the large and small islands of Miyakojima, Ishigakijima, and the Senkakus southwest of Okinawa) is an important passageway constrained by the island chain that the Chinese navy must break through to enter the oceans."[38] Guo Yadong of the Chinese Naval Studies Institute defended the transit of a ten-ship flotilla through the Ryukyus in 2010 on military grounds. Rapid advances in precision-guided weaponry, the need for training under meteorological and electromagnetic conditions, and the requirement to bolster logistics demand access to the high seas. Consequently, Guo claims, "the first island chain has already become the bottleneck that the Chinese navy's march to the deep blue must shatter."[39]

The Ryukyu island chain is also a major staging area for US military power in the Western Pacific. Shen Weilie of the Chinese National Defense University views Okinawa as the forward position of the US westward strategy in Asia.[40] He argues that cities like Shanghai, Hangzhou, and Xiamen are within striking distance of the island, while the Osumi and Miyako Straits could be monitored and blockaded from there. Chinese strategists have also been quite candid about the operational importance of this island perimeter to Japan during a cross-strait scenario. Aviation units deployed forward along the Ryukyu chain, contends Li Zhi, would play a critical part in contesting Chinese control of the air and sea.[41] Chinese analysts thus carefully track the military disposition of the Japanese Self-Defense Forces along the Ryukyus.[42]

LUZON STRAIT: A GAP IN THE CHAIN?

Chinese commentators devote the majority of their attention to the northern half of the first island chain, rarely describing the Philippine archipelago in ominous geostrategic terms. For one thing, economic centers and military bases on the coast do not face the distant southern portion of the island barrier. For another, Manila does not have the ability to challenge Chinese movements along the littorals. Most importantly, the Luzon Strait is one prominent passageway in and out of the South China Sea that appears promising to Chinese strategists. The Bashi Channel, separating Taiwan and the islands of Batanes Province of the Philippines, is one of the widest and deepest of the narrow seas found along the first island chain, with the waters southeast of the channel plunging more than three thousand feet.[43] Thus, it may be the key outlet for Beijing to break out of the China seas.

Owing to unique meteorological and oceanographic characteristics, the strait is particularly well suited for Chinese submarine operations. Atrocious weather conditions common to this area severely complicate airborne antisubmarine warfare operations.[44] Furthermore, the confluence of environmental factors peculiar to the Luzon Strait maximizes the efficacy of stealthy submarines. According to Du Pengcheng and Hu Chengjun of the Chinese Navy Submarine Academy, the Kuroshio Current, which flows through the Luzon Strait and passes the east coast of Taiwan, the Ryukyus, and Kyushu, is particularly conducive to sonar-reflecting undersea thermal layers that conceal submarines from ships and aircraft.[45] The layers are thick, wide, and stable year-round in the deep waters of Luzon.[46] Submarine commanders who know the local waters could exploit such conditions to breach the island chain undetected.

To Yu Fengliu, the strait is "a maritime area with extremely high economic, military, and political value worth its weight in gold, a nautical zone boasting important strategic meaning in the Western Pacific, and a channel for China to go past the first island chain worthy of close attention."[47] Because the Luzon Strait is the largest gap in the first island chain and presents a complex environment for the US Navy, Yu contends that Chinese air and naval forces could sortie through the strait without shore-based cover. His confidence speaks volumes about the strategic significance of the Sanya naval complex on Hainan Island.

THINKING ABOUT ISLAND CHAIN CAMPAIGNS

The voluminous literature on the first island chain reveals both a deeply felt insecurity about the seas and an urgent sense that China needs to acquire the wherewithal to control its surroundings. Chinese analysts believe that it is possible to mitigate the risk of encirclement. The call for more effective naval capabilities echoes the view that military power is the best antidote to Beijing's predicament. As PLA modernization progresses, Chinese leaders may gain confidence that the military could keep maritime communications open in a crisis and war.

The continued growth of Chinese sea power could loosen the maritime straitjacket. The feverish PLA investment in land-based precision-strike systems—designed in part to paralyze operations launched from bases along the first island chain—would complicate defense of the approaches to narrow seas. The long-awaited antiship ballistic missile, if it meets expectations, could allow China to control the seas from land. If these missiles can effectively keep warships out of Asian waters while destroying fighters before they take off from their bases, then China could suppress enemy movements to penetrate the island barrier.

With more and more-capable forces available to assert local sea and air control, Beijing could also apply pressure on multiple axes along the entire length of the first island chain. With enough assets on hand, it is conceivable that China would feint, staging a breakthrough at one location to distract an adversary while mounting its main effort elsewhere. The wider points of egress, including the Miyako and Luzon Straits, are the likely locations for diversionary tactics. For example, major sorties by the East Sea Fleet could draw attention to the Ryukyus, allowing the South Sea Fleet to support a breakout south of Taiwan. Whether US and allied forces could muster the assets to prevent such a breach in a timely way is open to debate.

On the other hand, PLA forces could mount a narrowly focused attack to open a corridor through the Ryukyus. Chinese forces might capture islands adjoining one or two straits through the archipelago. Prime candidates include Miyako and Ishigaki, which abut both the Miyako and Ishigaki Straits, the passages of choice for PLAN flotillas over recent years as evidenced by their deployment patterns. Missile-armed units on Miyako and Ishigaki could secure Chinese shipping in the straits. Occupying both islands would diversify operational planning.

Recent Chinese doctrine suggests that Chinese planners have closely studied the operational requirements for undertaking complex island-seizing operations. An impressive historical survey of amphibious campaigns of the late 1940s and 1950s analyzes nine examples of PLA assaults on offshore islands.[48] Another analysis contends that jointness, vertical envelopment, and beyond-the-horizon attacks will characterize future landing operations.[49]

The PLA volume *The Science of Campaigns* pays considerable attention to conducting amphibious landings under "informatized conditions." To successfully put military forces ashore, it exhorts the PLA to achieve sea control, air superiority, and information dominance.[50] The Nanjing Army Command College has issued a comprehensive assessment of island warfare across critical phases of a campaign. The contributors discuss at length defensive operations to consolidate PLA control over captured islands. For example, they forecast intervention of a powerful adversary (code for the United States) that could dislodge the Chinese military from its island positions. The authors detail antimine, antisubmarine, and anticarrier tactics to break a counter-blockade by such a third party.[51] Chinese strategists are clearly thinking in concrete operational terms about the implications of an island campaign.

An island campaign would serve multiple purposes by facilitating what the Pentagon calls antiaccess and area-denial measures to bar US reinforcements from the region while keeping forces in theater from areas like the Taiwan Strait. Operating between the first and second island chains, PLAN submarines and surface action groups could mount a defense in depth against US forces from Guam, Pearl Harbor, San Diego, Sasebo, or Yokosuka. Missile and torpedo attacks could inflict serious damage on resources needed to fight in the theater.

However, China's island chain strategy is not a risk-free proposition. The PLA confronts technologically advanced island nations that not only jealously guard their prerogatives at sea but also have honed skills and tactics to defend such privileges. Indeed, a determined and resourceful adversary could keep Chinese power-projection capabilities trapped behind the first island chain. One need look no further than the history of the Cold War. The Japanese Maritime Self-Defense Force effectively blocked chokepoints along the nation's archipelago with submarines, fixed- and rotary-wing antisubmarine warfare assets, and an elaborate undersea monitoring system, which bottled up Soviet submarines in the Sea of Japan.

A similar stranglehold could be applied to China. Japanese submarine and offensive mine campaigns along the Ryukyus could wreak havoc on Chinese seagoing forces seeking to seize some islands or hazard the main oceanic exits. If the PLA succeeded in putting forces ashore, allied airpower could raid Chinese positions to deny the invaders enough time to dig in. Shore-based assets, including mobile antiship and antiaircraft units deployed on those islands, could also contest PLA command of the air and sea in the East China Sea. To Tokyo's south, Taipei could adopt similar tactics to balk Beijing's strategy. Although a Chinese missile blitzkrieg would likely knock out Taiwan's capacity to control the skies and seas around the island, Taipei would not need to command the commons if its primary objective was to deny China a quick, decisive victory. Clever uses of fast attack craft, sea mines, truck-mounted cruise missiles, and the like could impose mounting costs on PLA forces operating near or over the island.

Such a hedgehog, or porcupine, strategy would see Japan and Taiwan settle into a defensive crouch. In effect, they would erect no-go zones along the Chinese frontier, replicating some of the PLA antiaccess and area-denial stratagems. By refusing to slug it out symmetrically, allied and friendly forces would expose Chinese forces to greater risk while preserving their main fighting units. Withholding operational successes from China would enable defenders to absorb and buy time to recover from the first PLA blows. Blunting any breakthroughs by Chinese forces along the first island chain would give way to efforts to keep the air and maritime corridors open for US reinforcements and follow-on counteroffensive operations.

Despite the resources that Beijing devotes to sea power, nothing is preordained about the Chinese ability to penetrate the first island chain. Even weaker members of the island barrier could trigger problems if they exploited their offshore positions. One implication is that Chinese nightmare scenarios about maritime confinement are not the products of overactive imaginations. If the island powers devise effective strategies and procure weapons to seal off PLA forces, then the worries expressed by Beijing are well founded. Another implication is that US allies should play on Chinese fears of entrapment by emphasizing the potential risks to its breakout plans that can shore up deterrence. In short, this seesaw contest for access will likely be a major feature of interstate competition along the East Asian littoral in the coming years.

• • •

Chinese literature on the first island chain demonstrates that well-articulated ideas on geography matter. To influence policy decisions in a competitive intellectual environment, Chinese analysts have structured their arguments about the island barrier in a manner that identifies the challenges confronting China with clarity and consistency. The concept of an island chain provides the focal point for developing sea power while sounding a clarion call for action. It is a vivid reminder that although many nations in a postmodern world are unaccustomed to thinking in geographic terms, some powers continue to employ the blunt language of geopolitics.

Globalization and interdependence have not deterred those inclined to see the world in geospatial terms. In the case of China, geography and history have fused, reinforcing perceptions of vulnerability to seaborne challenges. As some warfighting scenarios outlined above suggest, physical possession of real estate is still the final arbiter in international affairs, with steel hulls and airframes occupying the battlespace at sea. As long as the populations of the first island chain do not buckle under Chinese pressure, Beijing cannot assume that forcible passage through those islands is risk free.

Finally, strategy actually matters. In a contest between two or more living forces, Chinese efforts to break out of the island chain will stimulate countervailing reactions that hem in China. Shrewd tactics, effective use of resources, and maximum exploitation of geographic conditions will enable weaker powers to exact a high cost on prospective Chinese sea power that possesses quantitative and qualitative superiority. Strategy and operational art may prove as important as advanced technology for China and its neighbors along the East Asian littoral.

There is no way to know how much weight the geopolitical argument carries at the highest levels of decision making in Beijing. However, Chinese writings suggest that it would be imprudent to presume that geography plays no role in decision making. Some clearly do not like what they see on the map today. If the literature is an indication of the Chinese worldview, it behooves the United States and its allies to anticipate efforts that Beijing might take in the future to reshape its surroundings more to its liking.

NOTES

1. Alfred Thayer Mahan, *The Influence of Sea Power upon History, 1660–1793* (New York: Dover, 1987), 28.

2. See, for example, Du Chaoping, "How Much Influence Does the Island Chain Have on the Chinese Navy?" *Shipborne Weapons* 5 (2004): 37.

3. Liu Hong, "Geopolitical Consideration of the Taiwan Problem," *Journal of Yangzhou College of Education* 21 (December 2003): 48.

4. Sang Hong, "The Role of the Pacific and East Asian Littorals in Maritime Strategy," *Ocean World* 9 (2008): 43; John Foster Dulles, statement before the Senate Committee on Foreign Relations, 84th Cong., 1st Sess., February 7, 1955.

5. Chen Chungen and Jiang Sihai, "Taiwan: The Basis of US Geostrategy in the Asia-Pacific Region," *Journal of Jiangxi Institute of Education* 23 (August 2004): 24.

6. Huang Yingxu, "On the C Shaped Encirclement by the US," *Study Times* 154 (2010): 1.

7. For an early assessment of the first island chain, see You Ji and You Xu, "In Search of Blue-Water Power: The PLA Navy's Maritime Strategy in the 1990s," *Pacific Review* 4, no. 2 (1991): 137–49.

8. Liu Huaqing, *Selected Military Writings of Liu Huaqing* (Beijing: PLA Press, 2008), 528.

9. Ibid., 467.

10. The China National Knowledge Infrastructure contains one of the earliest uses in open-source literature of *first island chain* by the PLAN East Sea Fleet. See the abstract by Tian Renwu, "Field Observation and Analysis of Shengsi Port Attacked by Typhoon No. 8310," *Coastal Engineering* 2 (1984).

11. Liu Baoyin and Yang Xiaomei, *Island Chain Surrounding China—Oceanic Geography, Military Location, and Information Systems* (Beijing: Haiyang, 2003), 3–4.

12. Ibid., 17.

13. Ibid., 6–7.

14. Yu Kaijin, Li Guangsuo, and Cao Yongheng, "An Analysis of the Island Chain," *Ship and Boat* 5 (October 2006): 13.

15. Feng Liang and Duan Tingzhi, "Characteristics of China's Sea Geostrategic Security and Sea Security Strategy in the New Century," *China Military Science*, January 1, 2007, pp. 22–29.

16. Peng Guangqian and Ren Xiangqun, "Security Situation: What Is the Current Status?" *Global Military* 1 (2002): 8–9.

17. Wang Chuanyou, *On Maritime Frontier Security* (Beijing: Haiyang, 2007), 223.

18. Shao Yongling and Shi Yanhong, "The Fate of European Composite Land-Sea Powers in Modern History and Contemporary China's Choices," *World Economics and Politics* 10 (2000): 50.

19. Shen Weilie, "Thoughts on China's Future Geostrategy," *World Economics and Politics* 9 (2001): 73.

20. Li Yihu, "From Sea-Land Division to Sea-Land Integrated Planning: A Reexamination of China's Sea-Land Relations," *Contemporary International Relations* 8 (2007): 7.

21. Tang Yongsheng, "Actively Promote Westward Strategy," *Contemporary International Relations* 11 (2010): 19–20.

22. Liu Zhongmin, "Some Thoughts on Sea Power and the Dilemmas of Rising Great Powers," *World Economics and Politics* 12 (2007): 13–14.

23. Kong Xiaohui, "The Geostrategic Choices of China as a Composite Land-Sea Power," *Journal of University of International Relations* 2 (2008): 17.

24. Peng Guangqian and Yao Youzhi, *The Science of Military Strategy* (Beijing: Academy of Military Science, 2005), 443.

25. Wang Wei, *Geopolitics and China's National Security* (Beijing: Junshi Yiwen, 2009), 150.

26. Wang Chunyong and Lu Xue, "A Geostrategic Analysis of the Taiwan Question," *Journal of PLA University of Foreign Languages* 23 (May 2005): 114–15.

27. Zhang Shirong, "The Taiwan Question Is the Core Content of China's National Security in the Early 21st Century," *Journal of Yinchuan Municipal Party College* 39 (February 2006): 37.

28. Li Yaoxing, "The True Events of the Successful Passage Through the Taiwan Strait to Support Our Warships During the Paracels Sea Battle," *Party History Collection* 10 (2010): 46–48.

29. Liu and Yang, *Island Chain*, 52.

30. Li Jie, "The 'Island Chain' That Ties Up China," *Modern Ships* 7 (2001): 37.

31. Zhan Huayun, "Oceanic Exits: Strategic Passageways to the World," *Modern Navy*, April 2007, p. 28.

32. Zhu Tingchang, "On the Geostrategic Status of Taiwan," *Forum of World Economics and Politics* 3 (2001): 67.

33. Zhang Songfeng, "Sino-Japanese Relations in the Process of China's Peaceful Rise from a Geopolitical Viewpoint," *Modern Economic Information* 23 (2009): 215.

34. Liu and Yang, *Island Chain*, 17.

35. Ibid.

36. Feng Liang and Duan Tingzhi, "Japan's Oceanic Security Strategy: Historical Evolution and Actual Influence," *Forum of World Economics and Politics* 1 (January 2011): 78.

37. Editorial Board, "China's Sea Exit Problem Has Not Been Completely Resolved," *Water Transportation Digest* 11 (2004): 27.

38. Zhang Xiaowen, "The Strategic Intentions Behind and Influence of Recent Heightening Tensions in the Western Pacific by the United States," *Northeast Asia Forum* 93, no. 1 (2011): 55.

39. Guo Yadong, "China Must Resist the Noise of Threat Theory; Insist on Forging Blue-Water Navy," *Global Times*, May 5, 2010.

40. Shen Weilie, "Ryukyu, Island Chain, Great Power Strategy," *Leadership Digest* 5 (2006): 63.

41. Li Zhi, "Japanese and South Korean Aviation Units: Application to Naval Operations and Influence on China," *Shipborne Weapons* 12 (2007): 50.

42. See, for example, Gao Hui, "Japan's Military Deployment at Yonaguni Island and China's Maritime Security," *Naval and Merchant Ships* 9 (2009): 26–29.

43. Li Jie, "Taiwan Island and the Geography and the Climate in the Surrounding Seas," *Contemporary Military* 5 (May 2006): 70.

44. Liu and Yang, *Island Chain*, 96.

45. Du Pengcheng and Hu Chengjun, "Kuroshio East of Taiwan Island and Submarine Warfare," *Journal of Sichuan Ordnance* 31 (January 2010): 80.

46. Ying Nan, "The Oceanographic Environment of Taiwan's Sea Areas and Its Influence on Maritime Operations," *Contemporary Military* 7 (July 2006): 70.

47. Yu Fengliu, "The Best Sea Lane In and Out of the Pacific: Strait of Luzon," *Modern Navy*, May 2007, p. 20.

48. Wang Wei and Zhang Depeng, eds., *Cross-Sea Island Landings: Research on Operational Cases and Operational Methods* (Beijing: Military Science, 2002).

49. Ceng Sunan, Yu Kun, Sun Xihua, and Yan Hao, eds., *Amphibious Operations: History and Future* (Beijing: Military Science, 2001), 149.

50. Zhang Yuliang, ed., *The Science of Campaigns* (Beijing: National Defense University, 2006), 313.

51. Chen Xinmin, Shu Guocheng, and Luo Feng, eds., *Research on Island Operations* (Beijing: Military Science, 2002), 297–301.

4 MAHAN AND THE SOUTH CHINA SEA

James R. Holmes

WHAT WOULD THE fin de siècle theorist of sea power Alfred Thayer Mahan think about the strategic geography of the South China Sea? One thing is certain: he *would* think about it. He would draw comparisons with the Caribbean Sea, the Gulf of Mexico, and the Mediterranean Sea to tease out insights on strategic effectiveness in this semienclosed expanse. It is distressingly commonplace among contemporary strategists to bowdlerize Mahan, reducing his theories to advocating battles between swarms of armored dreadnoughts. Relegating him to the level of tactics and warships is a misrepresentation. For Mahan, as one noted historian declared, "sea power was the sum total of forces and factors, tools and geographical circumstances, which operated to gain command of the sea, to secure its use for oneself and to deny that use to the enemy."[1]

The instruments of naval warfare were important to the maritime equation, but far from the only factor. Mahan regarded sea power as interaction among domestic industry and foreign trade and commerce, commercial and naval shipping, and forward bases to support voyages by steam-powered vessels.[2] "Commercial value," he pointed out, "cannot be separated from military in sea strategy, for the greatest interest of the sea is commerce."[3] In contemporary terms, his intent was securing and maintaining political, military, and commercial access to regions such as East Asia. The starting point for the understanding of sea power was "the necessity to secure commerce, by political measures conducive to military, or naval, strength. This order is that of actual relative importance to the nation of the three elements—commercial, political, military."[4]

Mahan was acutely conscious of geography—more than any of the great strategic theorists except for the land-power scribe Antoine-Henri Jomini.[5] Indeed, some pundits pronounce Mahan to be a seagoing Baron Jomini.[6] For Mahan, studying geographic surroundings was a prerequisite for competitive maritime endeavors. He stated that geography underlies strategy.[7] Many rules of continental warfare mapped to the nautical milieu, which made the feats of Frederick the Great and Napoleon Bonaparte worthy objects of study. Mahan enjoyed citing the Napoleonic maxim that war is a business of positions, including four references in *Naval Strategy*.

When contemplating entry into an oceanic theater, makers of strategy must first survey the physical environment. Only by evaluating strategic features, determining which are critical, and integrating them with forces that shape events can a maritime power create an effective strategy. "In considering any theater of actual or possible war, or of a prospective battlefield," according to Mahan, "the first and most essential thing is to determine what position, or chain of positions, by their natural and inherent advantages affect control of the greatest part of it."[8] The decisions on where to deploy forces constituted major activities for any power that coveted access to far-flung expanses.[9] Geography represented the fixed setting of the dynamic, intensely interactive sphere of human enterprise within which maritime strategy unfolded.

Yet Mahan went beyond a general injunction to afford geography its due. Over the course of his career, he constructed a framework for analyzing the importance of strategic features such as seaports, islands, and narrow waterways. His first book explored the Gulf of Mexico and America's inland waters (1883). He returned to the same theme in "The Strategic Features of the Gulf of Mexico and the Caribbean Sea," an essay that was published in *Harper's Magazine* and reprinted in *The Interest of America in Sea Power, Present and Future* (1897), and in his rambling *Naval Strategy* (1911), which single-mindedly concentrates on finding similarities and differences between continental and maritime warfare.[10] Interestingly enough, his most influential work, *The Influence of Sea Power upon History*, contains the *least* geographic content, beyond an injunction that the extent and conformation of territory are important determinants of maritime power. This may explain the propensity of strategists to overlook this pivotal element of his writings.

Where did likely theaters of competition and conflict lie? Mahan cast this question in terms of power as well as purpose. He observed that certain regions of the world, "rich by nature and important both commercially and politically, but politically insecure, compel the attention and excite the jealousies of more powerful nations."[11] A combination of abundant natural resources, lively trade and commerce, and frail government—in short, a power vacuum in an economically important region—would beckon the inexorable gaze of foreign powers. Great-power wrangling over Manchuria and the Korean Peninsula offered examples for Mahan, writing after the Sino-Japanese War, Russo-Japanese War, and Japanese annexation of Korea.

How did Mahan estimate the strategic value of geographic positions? As noted before, he considered overseas naval stations one of the pillars of sea power. External powers had to be choosy about the sites they picked, lest they disperse forces too thinly and expose themselves to wartime defeat. Writing in the 1890s, Mahan proposed that "the strategic value of any position, be it body of land large or small, or a seaport, or a strait, depends, 1, upon [the] situation (with reference chiefly to communications), 2, upon its strength (inherent or acquired), and, 3, upon its resources (natural or stored)."[12] This simple framework retains its analytical power today, once the technological advances since the age of Mahan are taken into account.

In maritime strategy as in real estate, the key advantage for naval strategists was location. Prospective bases should lie along strategic lines to be worth occupying, regardless of their innate strength or resources. Harbors astride critical sea lines of communication were ideal. Proximity to friendly seaports was another benefit, allowing mutual support in wartime. The fleet detachments could readily combine for defensive or offensive action. Proximity to hostile naval stations gave squadrons the prerogative to watch or interdict enemy movement. Gibraltar would be worthless as a naval station—notwithstanding its formidable natural defenses—if it were located in an expanse devoid of merchant and naval traffic.[13] "Strength and resources," observed Mahan, "may be artificially supplied or increased, but it passes the power of man to move a port which lies outside the limits of strategic effect."[14] Terrain could be improved to a degree, although resources could arrive overland or oversea. Sites were fixed.

Seaports needed military strength to both resist assault and project force outward. Natural defenses constituted a formidable advantage. Cliffs overlooking the seaward approaches made amphibious landings a daunting prospect but enabled defenders to rain fire down on attacking fleets from high ground. Similarly the narrow mouth of a harbor permitted overlapping gunfire from all sides. If a port lacked the innate capacity to withstand attack, a navy must undertake works to fortify it. A squadron stationed at an outpost able to protect itself could roam the seas confidently and execute assigned missions without fear of losing landward refuge.

"Resources (natural or stored)" meant a supply of needed goods for the residents of the port, allowing shipyards to repair, rearm, and reprovision visiting merchantmen and warships. Food, fuel, shipbuilding materials, and munitions were only some of the goods. Large islands and harbors with ample backcountry often abounded with resources. Other sites had to rely on shipments of critical goods. Despite its imposing defenses, for example, Gibraltar could not sustain itself for long without maritime trade.[15] Its relationship with the Royal Navy was symbiotic. Men-of-war based at Gibraltar could dictate the terms of access to the Mediterranean, but the base could not survive unless Britain ruled the waves, guaranteeing a steady flow of supplies.

Mahan warned against measuring the utility of any locale apart from the surrounding area. This was especially true in the cramped confines of America's Mediterranean. The arrangement of the Gulf of Mexico and islands in the Caribbean created a nearly continuous barrier. Cuba, Santo Domingo, and Puerto Rico constituted the principal obstacles. Narrow waters corralled shipping to or from the Isthmus of Panama into three principal routes. The Yucatan Channel passed to the west of Cuba, and the Windward Passage separated the eastern tip of Cuba from Haiti. Mahan therefore concluded, "Cuba is as surely the key to the Gulf of Mexico as Gibraltar is to the Mediterranean."[16]

Cuba also featured long coastlines, multiple harbors, and indigenous resources. Defenders could resupply Havana and Santiago overland, defying the efforts of a blockading fleet. Best of all from the American standpoint, the United States had won basing rights at Guantanamo Bay by virtue of its defeat of Spain in 1898. It was ascendant over Royal Navy squadrons based at Jamaica to the south. A third major sea line of communication, the Anegada Passage, lay to the east of Puerto Rico, another prize of war captured from Spain.[17] Together these island holdings empowered the United States to mount a forward defense of its coastline in the Gulf of Mexico, providing a central position for radiating power southward toward the isthmus.

The interdependence among sites as varied as Pensacola, Key West, and Guantanamo Bay complicated geostrategic calculations. Some sites were "overshadowed by others so near and so strong as practically to embrace them."[18] When evaluating the comparative merits of Jamaica and Cuba, for instance, he pointed out that Jamaica flanks all lines of communications.

Judged entirely in terms of geographic position, the British-held island possessed the greatest potential of any geostrategic asset in the Caribbean Sea. Yet it was deficient in resources and dependent on maritime trade. Cuba overshadowed Jamaica, controlling all sea communications with the lesser island. Only a fleet superior to an adversary based in Cuba would be capable of shielding Jamaica from isolation and being starved of critical supplies. It took a dominant navy to imbue Jamaica with the full value it held in abstract calculations.[19]

Mahan elaborated on his position-strength-resources template in *Naval Strategy*, extending it to straits and other confined waterways. "The military importance of such passages or defiles depends not only upon the geographical position," he wrote, "but also upon their width, length, and difficulty." Specifically, a strait was a strategic point whose value depended on its situation on the nautical chart; on its "strength, which may be defined to consist in the obstacles it puts in the way of an assailant and the consequent advantages to the holder." What is more, it relied on "resources or advantages, such as the facility it gives the possessor for reaching a certain point," or shortening distances from seaport to seaport.[20]

As in his analysis of bases, Mahan cautioned against appraising narrow seas in a vacuum. When fixing the value of a passage it was crucial to weigh the availability of nearby alternatives. "If so situated that a long circuit is imposed upon the belligerent who is deprived of its use, its value is enhanced." If a strait or passage constituted "the only close link between two bodies of water, or two naval stations," its importance was further magnified. Finally, the underwater conformation of a narrow sea must be taken into account. Hydrographic conditions such as convoluted channels, depth of water, and underwater hazards to navigation influenced the offensive and defensive potential of any passage.[21]

Finally, Mahan noted in passing that "a certain regard must be had to political conditions, which may be said to a great extent to neutralize some positions." Social or political upheaval in the surrounding country could work against or even negate the value of a site by undercutting its defensibility or impoverishing its stock of resources. Mahan dismissed Haiti as a base for that reason. The island's constant revolutionary upheaval, or sociopolitical nothingness, rendered it little more than an inert obstacle to US maritime strategy.[22] Such comments about the political, social, and cultural context have the feel of an afterthought for Mahan. But he did acknowledge that there were indices of strategic value apart from location, terrain, and sustenance for bases. Sizing up the local culture could be just as important for planners.

A MAHANIAN SURVEY

The South China Sea presents operational surroundings that are at once more and less hospitable for navies than other enclosed expanses of comparable size. It is wider and more vacant than the Mediterranean or Gulf-Caribbean combine, facilitating free passage for commercial and naval vessels while allowing naval task forces ample room to maneuver. No obstacles comparable to the Italian peninsula constrict navigation. And no island-chain barrier comparable to the Cuba–Santo Domingo–Puerto Rico line funnels shipping bound for the Malacca Strait, the main outlet to waters beyond, through focal points that can be guarded by watchful forces.

Indeed, only a handful of mostly small islands and reefs—the Spratlys to the south, the Paracels to the north—break up the largely featureless maritime plain that separates Vietnam from the Philippines along the east–west axis and Hong Kong from Borneo from north to south. The Spratly and Paracel Islands command enviable geographic positions while featuring next to nothing by the standards of strength and resources. Many are uninhabited and at best could play host to small bodies of troops equipped with antiship cruise missiles, providing the occupying power a sea-denial option vis-à-vis passing merchant or naval traffic.

Accordingly, no aspiring nautical hegemon—China being the most likely candidate—could easily mount a forward presence comparable to the US central position in Cuba following 1898. There is no Cuba, Puerto Rico, or St. Thomas from which to stage forward operations. Nor are there counterparts to Gibraltar or to the other Mediterranean outposts held by Great Britain during its maritime heyday. Hainan Island, home to a submarine base at Sanya, extends China's seaward reach, but only by some 140 miles from the coastline of China. Beyond that significant exception, positions on the periphery command more value than they did in Mahan's Gulf and Caribbean. Cam Ranh Bay, a US port built during the Vietnam War, constitutes a particularly attractive harbor astride the eastern approaches to the Malacca Strait. Changi, in Singapore, can berth US carriers as well as smaller craft. But with few options in the southern reaches of the South China Sea, naval forces capable of at-sea replenishment—an indispensable prerequisite for sustained operations on the high seas—will be at a premium in any maritime competition in Southeast Asia. This may well account for Beijing's feverish quest for aircraft carriers, the best apparent mobile substitute for land-based airpower.

The frontiers of the South China Sea bear closer resemblance to the Gulf and Caribbean than to the Mediterranean Sea. The latter is bordered entirely by continental landmasses, whereas the former is bordered by Southeast Asia to the north and west and by Taiwan and the Philippine and Indonesian archipelagoes to the east and south, in a massive arc sweeping from the Taiwan to the Malacca Strait. Its borders are more permeable than those of the Mediterranean. Alternatives to the Malacca Strait—namely, the Lombok and Sunda Straits—pass through the Indonesian archipelago along the southern tier. Similarly, Mahan emphasized that the Antilles offered few impediments to shipping despite auspicious positioning on the map.[23] Indeed, the southeastern fringes of the Caribbean Sea verge on being open sea. When armed and fortified, by contrast, the Philippines and Indonesia are better positioned to regulate the sea-lanes through or near their waters. A combination of small craft, land-based antiship missiles, and underwater mines would give them significant leverage in wartime.

Contemporary strategists are unable to simplify the geometry of the South China Sea as neatly as Mahan simplified that of the Gulf and Caribbean. The naval historian scribed a triangle over the region that contained the all-important geostrategic features of this inland sea. A line connecting New Orleans with Colón, Panama, formed one side. A second leg originated at Pensacola and ran through and beyond St. Thomas. The final leg started at Colón and ran through Cartagena and Curaçao, intersecting with the Pensacola–St. Thomas leg to the east of Martinique. Mahan believed that it was possible to concentrate analytical energies for two reasons. First, application of the position-strength-resources template failed to reveal any seaports of consequence between New Orleans and the northern tip of the Yucatan Peninsula. Second, Mexico was stable but it had little naval power. By this logic strategists could effectively disregard the mostly desolate coastline west of the Mississippi Delta. It was inert from the standpoint of sea power.[24]

It would be perilous to discount the potential of Southeast Asian nations in the manner that Mahan discounted Mexico. China certainly boasts the most potential for sea power of any littoral state. But unlike Latin America in Mahan's day, even the lesser nations of Southeast Asia can influence their maritime environs. Some, like Singapore, can deploy small but first-rate navies. Singaporean sailors are known for their skill and élan, and they have

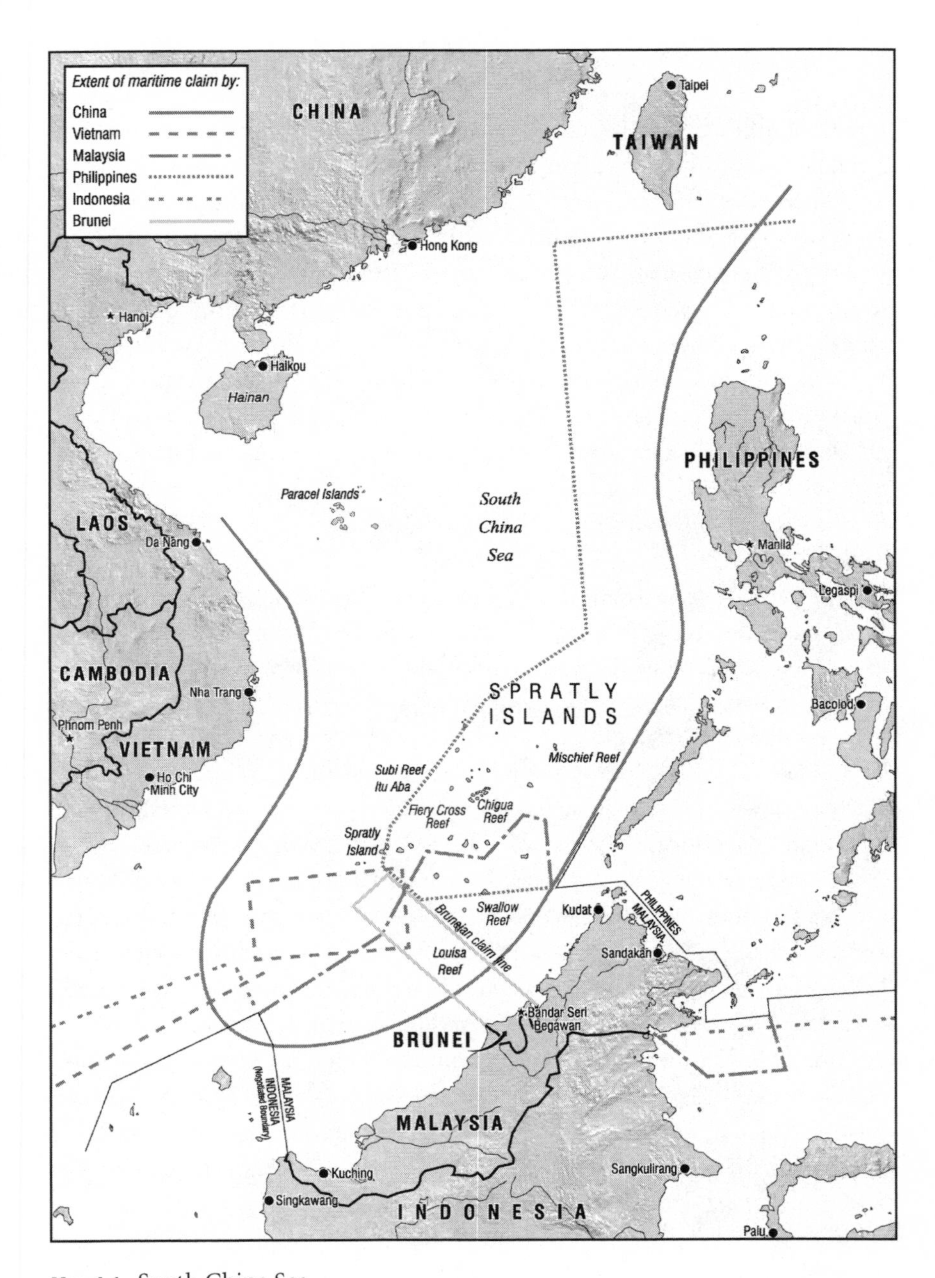

Map 4.1. South China Sea.

This body of water is designated as a marginal sea, meaning that it is partially enclosed but open to an ocean. Geopolitically, marginal seas are considered territorial waters. The area of the South China Sea is approximately 1.4 million square miles and borders mainland China, Taiwan, Vietnam, the Philippines, Brunei, Malaysia, and Singapore. These waters encompass 250 islands, shoals, atolls, and reefs and include the Paracel and Spratly Islands.

effective weapons systems. Others, notably Vietnam, have set out to field viable maritime forces of their own. Hanoi intends to acquire six top-flight *Kilo*-class diesel submarines from Russia, complementing its fleet with sea-denial measures vis-à-vis the People's Liberation Army Navy (PLAN). A Vietnamese *Kilo* squadron could oppose efforts by China to exert primacy in regional waters while complicating PLAN attempts to exploit the potential of its submarine base on Hainan Island. Indonesia also has announced plans to increase its maritime reach. Even the Philippines, despite its negligible defense budget, has options in the form of a long-standing mutual defense pact with the United States and a history of playing host to powerful US naval and air forces.

No appraisal of the South China Sea would be complete without a few words on the value of Taiwan. The parallels between Taiwan and the islands Mahan appraised are inexact at best, but the points of difference are revealing. Taiwan is something like Cuba by Mahanian metrics. In terms of position, it abuts crucial sea-lanes running north and south along the Asian seaboard, through which merchantmen carry raw materials and finished goods to and from northeast Asian economies. Its northern tip faces the southernmost links in the Japanese Ryukyu island chain. Its southern tip, moreover, overlooks east–west transit between the South China Sea and the Western Pacific through the Luzon Strait. Apart from its fortuitous positioning, the island is sizable and well stocked with certain resources, notably foodstuffs, although it depends on outside supplies of fuel. Whoever controls the island of Taiwan has the ability to move forces overland, helping defenders evade a blockade. Seaports of various sizes and shapes dot its lengthy coastline, providing bases for fleets of small craft as well as a navy.[25]

On the other hand, Taiwan is much like Jamaica in Mahan's formulation. Although the island flanks key sea lines of communications, the Chinese coastline overshadows the island. Only superior forces based on Taiwan would let the island reach its potential as a geostrategic asset in the face of Chinese enmity. Should mainland China restore its rule over Taiwan, the island would come to resemble Key West, with enormous offensive and defensive potential.[26] This new-old asset would extend the Chinese reach eastward into the Western Pacific; turn the southern flanks of Japan and South Korea, giving Beijing newfound geostrategic leverage; and enable PLAN warships to command the northern rim of the South China Sea and also project power to the Luzon Strait and elsewhere in the northern reaches of that expanse.

Thus Chinese forces stationed on the island could shield the mainland from prospective adversaries such as the United States, control the seaborne communications of competitors in East Asia, and be guaranteed access through the Luzon Strait while denying it to adversaries. Small wonder that Admiral Ernest King, the chief of naval operations during World War II, affirmed that whoever controlled Formosa enjoyed the freedom to put the cork in the bottle of the South China Sea.[27] Such observations explain why the United States placed such value in Taiwan during the Cold War. This unsinkable carrier and submarine tender helped anchor containment policy vis-à-vis the Soviet Union and China.[28] This approach to maritime strategy animates Beijing today, imparting urgency to its policy toward Taipei.

Finally, there is a critical difference between the South China Sea and its precursors—the Gulf-Caribbean combine and the Mediterranean Sea—in an external line of communication that bypasses the region. Specifically, forces based in Australia to the south can move from one side to the other between the Indian Ocean and Western Pacific, without venturing into the South China Sea. Therefore, Australia holds another central position in the grand Indo-Pacific region, offering an alternative to Singapore and the Malacca Strait. Recent moves to base US forces in strategically located sites such as Darwin on the northern Australian coast recognize and take advantage of this convenient geostrategic reality.

• • •

The South China Sea represents a massive maritime crossroads of enormous potential value for seafaring nations although presenting few opportunities for permanent forward bases. It will prove difficult for any would-be hegemon to command, even for a coastal nation like China that is replete with maritime potential. Accordingly, a blue-water fleet capable of projecting power throughout the region will be required by any nation that covets maritime command. Moreover, combat logistics vessels will be necessary to replenish this fleet so that it can remain on station for the weeks or months that sustained campaigns demand.

Beijing has taken on an imposing slate of commitments along China's nautical periphery, ranging from managing events from the Korean Peninsula to managing maritime security in the Malacca Strait and waters of the extreme southwest. Moreover, Chinese naval modernization remains a work in progress, which means that decisions to concentrate resources in Southeast

Asia will come at the expense of pursuing interests elsewhere. Alfred Thayer Mahan would entertain doubts about the capacity of the People's Republic of China or any other seafaring nation to enforce its will on Southeast Asia any time soon.[29]

NOTES

1. William E. Livezey, *Mahan on Sea Power* (Norman: University of Oklahoma Press, 1947), 277.

2. Alfred Thayer Mahan, *The Influence of Sea Power upon History, 1660–1783* (Boston: Little, Brown, 1890), 71.

3. Alfred Thayer Mahan, *Naval Strategy Compared and Contrasted with the Principles and Practice of Military Operations on Land* (Boston: Little, Brown, 1915), 302.

4. Alfred Thayer Mahan, *Retrospect and Prospect: Studies in International Relations, Naval and Political* (Boston: Little, Brown, 1902), 246.

5. Mahan, *Naval Strategy*, 107.

6. See, for example, Brian R. Sullivan, "Mahan's Blindness and Brilliance," *Joint Force Quarterly* 21 (Spring 1999): 115; and J. Mohan Malik, "The Evolution of Strategic Thought," in *Contemporary Security and Strategy*, ed. Craig A. Snyder (New York: Routledge, 1999), 36.

7. Mahan, *Naval Strategy*, 319.

8. Ibid., 22.

9. Ibid., 235–36.

10. Alfred Thayer Mahan, *The Gulf and Inland Waters: The Navy in the Civil War* (New York: Charles Scribner's Sons, 1883); Alfred Thayer Mahan, *The Interest of America in Sea Power, Present and Future* (Boston: Little, Brown, 1897); Mahan, *Naval Strategy*.

11. Mahan, *Naval Strategy*, 306.

12. Mahan, *Interest of America in Sea Power*, 283.

13. Mahan, *Naval Strategy*, 132.

14. Mahan, *Interest of America in Sea Power*, 283.

15. Mahan, *Naval Strategy*, 132–33.

16. Ibid., 347.

17. Mahan, *Interest of America in Sea Power*, 270.

18. Mahan, *Naval Strategy*, 380–82.

19. Ibid.

20. Ibid., 309–10.

21. Ibid.

22. Ibid., 346.

23. Ibid., 355–56, 364–65.

24. Ibid., 311–13.

25. "On the Cuban coast," declared Mahan, "there are so great a number of harbors that there can be no doubt of finding such as shall be in all ways fit for intermediate harbors, of refuge or for small cruisers." Mahan, *Naval Strategy*, 335. See also James R. Holmes and Toshi Yoshihara, *Defending the Strait: Taiwan's Naval Strategy in the 21st Century* (Washington, DC: Jamestown Foundation, 2010).

26. Mahan, *Naval Strategy*, 316.

27. Samuel Eliot Morison, *The Two-Ocean War: A Short History of the United States Navy in the Second World War* (Annapolis, MD: Naval Institute Press, 1963), 476.

28. See Wang Weixing, "Who Is the One That Wants to Push Taiwan into War?" *Jiefangjun Bao* [PLA Daily], March 15, 2000; Bi Lei, "Sending an Additional Aircraft Carrier and Stationing Massive Forces: The US Military's Adjustment of Its Strategic Disposition in the Asia-Pacific Region," *Renmin Wang* [People's Daily], August 23, 2004. Both articles quote remarks by General Douglas MacArthur that Taiwan is an unsinkable carrier and submarine tender (a metaphor apparently borrowed from the *New York Times*) and that its loss would pierce the American protective screen in the Western Pacific.

29. For an overview of dispersal, see Toshi Yoshihara and James R. Holmes, "Can China Defend a 'Core Interest' in the South China Sea?" *Washington Quarterly* 34 (Spring 2011): 45–59.

5 THE US ALLIANCE STRUCTURE IN ASIA

Michael R. Auslin

FOR MORE THAN SIX decades the alliance structure formed by the United States in the Asia-Pacific region weathered the Cold War, economic crises, and the War on Terror. Today that structure is faced with the ascendancy of China to regional prominence and the relative decline in the power of the United States and its allies in Asia. It is time to reassess the balance of power in Asia to determine whether existing alliances there are capable of meeting US strategic objectives. Challenges that arise over the next decade will involve minimizing disruptions from a shift in power and will require maintaining US presence and ensuring alliances are able to shape future events. Building on the alliance structure will promote Asian regional stability and preserve international security as long as the United States marshals the will and necessary resources.

DURABLE PARTNERSHIPS

The global alliance structure arose following the upheaval of World War II. Within a few years, Washington became the driving force behind international organizations and commitments that have lasted for more than six decades. This foreign policy strategy rested on political, military, and economic pillars derived from the Four Freedoms espoused by Franklin Roosevelt in 1941 and the liberal internationalism of Woodrow Wilson. In what quickly evolved into a global ideological contest for power, the goals of US foreign and defense policy became increasingly indistinguishable from the values spread by its international leadership.

On one level the United States fostered organizations such as the United Nations and the World Bank that were open to all nations to promote a new international order. A second level consisted of regional defense pacts like the North Atlantic, Southeast Asia, and Central Treaty Organizations. Finally, a third level comprised political, economic, and cultural activities; covert paramilitary operations; and military interventions to blunt support from the Soviet bloc to client states in the developing world. Given that agenda, the United States occasionally entered into partnerships with authoritarian regimes out of political expediency to maintain the balance of power, which complemented its broader national security objectives.

The Asia-Pacific alliance structure differs noticeably from collective security arrangements such as the Atlantic Alliance, also called the North Atlantic Treaty Organization (NATO). In Western Europe, a large multinational organization was created on the basis of common values and shared interests. In Asia, by contrast, bilateral security pacts were concluded with Australia and New Zealand (1951), the Philippines (1952), South Korea (1953), Thailand (1954), and Japan (1960) that obligated the United States to support those nations if land or seaborne invasions were remotely possible.

Beginning with the postwar occupation of Japan, the United States maintained a presence that soon led to permanent security agreements. Once the Korean War broke out, it became clear that disengagement was impossible. In the ensuing period of decolonization, no thought was given to creating new security organizations in Asia, particularly because the strategic focus remained on Europe.[1] Therefore, bilateral alliances were the most realistic means of protecting US interests and ensuring that key partners such as Japan and Korea did not fall into Communist hands. These arrangements formed a *hub-and-spoke* structure, with nations connected to Washington though not to one another. Moreover, these arrangements committed the United States to defend its partners. To meet its security obligations, the United States based land, sea, and air forces in three allied nations: Japan, the Philippines, and South Korea.

US forces in South Korea were committed primarily to defend against the regime in the north once a cease-fire was reached in 1953. In both Japan and the Philippines, however, the directive of American troops was broadly focused on regional stability. From their northern and southern bastions, the US Navy and US Air Force maintained a constant presence in East Asia. Naval bases at Yokosuka in Japan, Subic Bay in the Philippines, Kadena Air Base on Okinawa, and Clark Air Base in the Philippines were the largest American facilities in the region. These outposts supported combat operations in

Vietnam and kept tabs on Soviet forces, enabling the United States to project its power during the Cold War.

But as the Soviet Union developed its ballistic missile forces during the 1970s and 1980s, traditional security guarantees were no longer viable, and extended deterrence, whereby the United States pledged to launch a nuclear response to a nuclear missile attack on its allies, became a central feature of Asian security alliances. To a certain extent this changed the role of US conventional forces, which began to focus more on limited defense of the region, including any attempt by mainland China to invade Taiwan or start another Korean war.

As the Cold War ended, the US alliance structure was faced with new strategic challenges. The rationale for stationing tens of thousands of troops and millions of dollars in hardware far from the continental United States seemed less persuasive. The Russian Navy was rusting in Vladivostok and, although the Kremlin deployed thousands of nuclear missiles, tension declined between Moscow and Washington. In the last several decades, there have been fears of nuclear conflict but little concern over a traditional war in Asia other than in Korea. Yet even today, Japan provides the home port for the only forward-based carrier that maintains the credibility of American presence across the vast expanses of the Pacific.

As the United States evaluated the post–Cold War world, new concern over Asian stability emerged with the rise of China. After the 1970s and decades of the Maoist regime, US strategy sought to integrate China into the global order and shape political relations to take advantage of the Sino-Soviet split.[2] But even with the economic reforms of Deng Xiaoping in the late 1980s and 1990s, Beijing continued to suppress liberal influence and undertook a military buildup to acquire advanced weapons that seemed designed for offensive purposes. Chinese agents stole ballistic missile technology from the United States in the 1990s and began buying submarines, surface warships, and fighters. The modernization of the People's Liberation Army (PLA) has been increasingly directed at US and allied forces in the region with the purchase of capabilities that contribute to a strategy known as *antiaccess and area denial.*

From both Asian and American perspectives, it was worrying that the Chinese approach to foreign affairs seemed to harden as the nation strengthened. Despite diplomatic efforts to create a unique G2 strategic partnership, Beijing and Washington found themselves increasingly at odds with one another. China argued that its assertiveness was driven by the security environment and interests of a rising power.[3] Washington, however, regarded certain actions by Beijing as threats to regional stability. From deploying more than a thousand ballistic missiles on the Taiwan Strait to badgering the nations of East Asia over disputed territories and exclusive economic zones, China seemed intent on translating wealth into regional dominance. Therefore, within a decade of the Cold War, the United States redefined its policy toward Asia to stabilize the changing international environment and reaffirm its commitments to allies. This led Beijing to denounce Washington for initiating a policy of containment, which invoked memories of the earlier standoff between the United States and the Soviet Union.

THE ALLIANCE STRUCTURE

By historical standards, the changes in Asian economic and military conditions have been rapid. Within a few decades, China moved from fielding antiquated, largely territorial armed forces to deploying advanced weapons and developing power-projection capabilities. North Korea, for its part, has made itself a de facto nuclear power with growing ballistic missile capabilities.[4] In response, nations across the region are seeking to bolster their defense postures as they continue to depend on the United States. Many Asian navies are buying submarines or building their own, and richer nations are buying advanced fighters. Moreover, antiship cruise missiles are becoming more prevalent, with North Korea reportedly testing such a missile in 2011.[5] Other short-range offensive systems are being pursued. For example, after its warship *Cheonan* was sunk by North Korea in 2010, South Korea began producing ship-to-submarine Hong Sangeo missiles.[6] In addition, Russia and India have been developing the BrahMos supersonic cruise missile, and India has reportedly begun negotiating sales of military technology to Vietnam.[7]

Such change seems particularly disruptive in Asia, which lacks the security arrangements and relationships of trust similar to those developed in Western Europe following World War II. Moreover, memories of Japanese imperialism continue to roil some nations in the region. Japan has been prevented from forging closer ties with South Korea because of its colonial legacy, and China continually demands that Japan apologize for its wartime atrocities. Equally important, distrust of China has thwarted establishment of broader regional institutions.

The United States has played the role of balancer in the regional system without seeking territory or directly influencing governments there. However, its balancing act has not been neutral or disinterested. Instead, its policies have sought to protect US national interests in concert with democratic nations, though that has included relations with authoritarian states in the past. Looking at its long-term strategy, the US approach toward China is more intelligible, with Washington working to integrate Beijing into a global political and economic order, with the proviso that increased Chinese power should not threaten the region.

The rise of China has resulted in a situation different from the Cold War, when US allies enjoyed little if any trade with the Soviet Union and often only perfunctory diplomatic relations. Moreover, there was a clear divide between the Soviet bloc in Asia and US regional interests. Today all nations in the region have extensive trade relations with China, which is the biggest trading partner for many US allies. In the case of Japan, for example, Chinese exports rose to more than 13 billion yen in 2011 while US exports amount to under 6 billion yen.[8] The nations of the region maintain diplomatic relations with the United States and China and participate in initiatives with the Asia-Pacific Economic Cooperation Forum, the Association of Southeast Asian Nations, and other multilateral organizations. This engagement has caused some Asian nations to become ambivalent over whether China is a partner or a threat.

Many Asian nations depend on China for economic growth and thus are loath to antagonize Beijing. At the same time, they fear that modernization of the Chinese military will threaten their national interests. Moreover, they are concerned about the brashness of China in diplomatic rows over contested territories. In particular, the willingness of Beijing to coerce smaller nations raises the fear that a more powerful China will be less likely to promote international norms of conduct. Changes in the geopolitical balance have led the United States to consider two possible threats to its alliances: conflict over territorial claims and disruption in the commons.

TERRITORIAL DISPUTES

The first threat to the US alliance system is the potential for direct conflict among Asian nations. The lack of success in resolving territorial disputes has resulted in regional tension. Nearly every Asian nation lays claim to contested territory, including the Korean Peninsula, the sovereignty of Taiwan, Chinese control of Tibet, sovereignty of the Senkaku/Diaoyu Islands (claimed by China and Taiwan and administered by Japan), control of the Spratly Islands (claimed by Brunei, China, Malaysia, the Philippines, Taiwan, and

Vietnam), control of the Paracel Islands (claimed by Vietnam and Taiwan but occupied by China), sovereignty of the Takeshima/Dokdo Islands (claimed by Japan and North Korea and controlled by South Korea), the sovereignty of the Kurile Islands (claimed by Japan and under Russian jurisdiction), and control of Arunachal Pradesh by China (though claimed by India). Although such claims resemble disputes in Europe during the nineteenth century, with nationalism and insecurity driving military expansion, particularly of the naval forces, the disputes in Asia are played out on a dispersed maritime tableau and frequently involve old disagreements over archipelagic or island chain boundaries.

The possibility of conflict over these territories has been increasing in recent years, spurred by claims in the East and South China Seas. Incidents over fishing rights in exclusive economic zones have pitted China against its neighbors. Recently, Chinese patrol craft intervened in the arrest of fishing boats and threatened law enforcement vessels of other nations. Such actions caused a crisis in 2010 when Beijing retaliated against Tokyo for arresting a fishing boat captain near the Senkakus.[9] Subsequently, Japan nationalized several islands, resulting in confrontations between Japanese Coast Guard and Chinese maritime patrol vessels.

Beijing has also begun to send remotely piloted aircraft and patrol aircraft into or near the airspace of the Senkakus. In Southeast Asia, Vietnam has especially borne the brunt of Chinese pressure over the exploration of oil fields under the South China Sea by interfering with drilling ships and threatening foreign companies that collaborate with Vietnamese in oil ventures.[10] The Philippines has clashed frequently enough with China over the Spratly Islands that Philippines president Benigno Aquino has publicly called for help in resisting Chinese coercion.[11] Manila's and Hanoi's complaints of intimidation by China of their vessels and the incursion of Chinese fishermen into contested areas like the Scarborough Shoal have raised tensions throughout the region.

Japan also has territorial disputes with other nations and has dispatched both Coast Guard and Maritime Self-Defense Force vessels to the Senkaku and Takeshima Islands during times of tension with Taiwan and South Korea, respectively. South Korean nationalists have demonstrated against the Japanese claims on Takeshima, even going as far as acts of self-immolation.[12] Other nations also have clashed over the disputed territories. In 2009, for

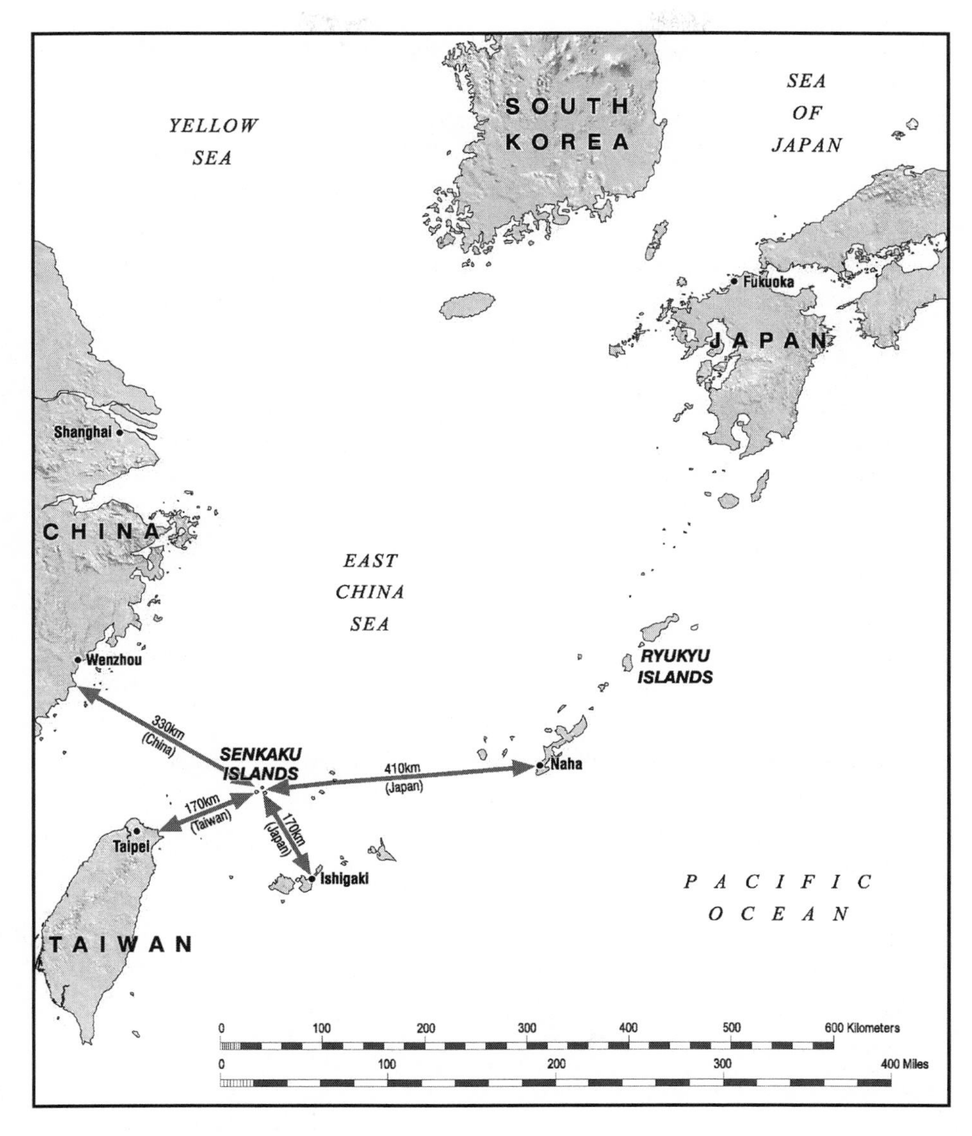

Map 5.1. Senkaku Islands.
This archipelago comprises five uninhabited islands and three reefs known as the Senkaku Islands by the Japanese and the Diaoyu (Tiao Yu Tai) Islands by the Chinese. Administered by Japan as part of Okinawa Prefecture, these islands are also claimed by both China and Taiwan.

example, naval forces from Indonesia and Malaysia faced off in an argument over oil-rich waters off Borneo.[13]

The United States does not have a direct role in these territorial disputes save for a pledge to defend its treaty allies. For the most part, US warships do not intervene in maritime face-offs; however, in two recent cases, allies sought clarification from the United States on their mutual security arrangements. In the first case, which occurred during the Senkakus crisis, Tokyo asked Washington whether Article V of their mutual assistance treaty covered the islands. The Japanese press reported that Secretary of State Hillary Clinton reassured the Japanese foreign minister that it applied, but official US statements were less unequivocal.[14] In a second case, in 2011, Manila announced that it had reminded Washington of their mutual defense treaty because of disputes with China over oil exploration and fishing rights in the South China Sea.[15]

Such incidents reveal how the alliance structure, forged for conventional warfare between nations, is being put under pressure to respond to incidents over exclusive economic zones and isolated islets with little overt strategic value. Washington has long urged its treaty partners to provide more for their own defense but now faces two problems: no ally feels strong enough to deal with China alone, and the regional military buildup may make conflicts more likely when diplomacy proves insufficient to address incursions into disputed waters.

ACCESS TO THE COMMONS

The second major threat to the US alliance system comes from the possibility of disruption and instability in the regional commons. Over the past decade, as China has developed its maritime and air capabilities and pursued its territorial claims with greater vigor, US strategy has shifted its focus to the risk of assertive naval and air power on freedom of navigation and access to the global commons. Bilateral alliances are being stretched to offer security assurances in the event of a possible regional crisis or an intra-Asian bilateral conflict.

The wider Asia-Pacific region stretches from the coast of the Indian subcontinent through Southeast Asia and Oceania to the Chinese littoral, Korean Peninsula, and Japan, ending in the Arctic routes of the Russian Far East. It connects the Western and Eastern Hemispheres and, more particularly, the energy-rich Middle East with East Asian markets. It also contains the most vital sea-lanes in the world, including the Lombok, Malacca, and Sunda

Straits through which more than seventy thousand ships pass every year and almost half of all global exports. Within the region, the Malacca Strait and lower South China Sea serve as a hinge between the Middle East and Indian Ocean region and the highly developed northwestern Pacific Rim.[16]

The growth in global production and consumption in the last half century of globalization was made possible in no small part by the economic vitality of the Asia-Pacific region. The free flow of goods facilitated economic expansion along with that of information networks and exchange of peoples. Disruption in this system could have serious repercussions.

Yet stability is as much a matter of mind as it is of objective reality. The more that nations are concerned over access to the global commons and suspect growing threats to their ability to participate in the global economy, the more likely they are to protect those interests. Hence, on one level, this results in increased military spending. On another level, it may lead nations to act in ways that would protect their national interests but that would also increase regional insecurity that could threaten the global commons, which represents the nexus of US alliances and regional stability. As a result, safeguarding freedom of navigation in major trade routes and in contested waters, such as those near the Spratly Islands in the South China Sea, has become a major concern for the United States.

Territorial claims can lead strong nations to exert pressure on weaker ones by establishing de facto zones of influence or no-go zones. Thus far, only Beijing has used such tactics. In 2011 the PLA Navy (PLAN) challenged an Indian vessel departing Vietnam by claiming it was operating in Chinese waters. An earlier incident, in 2009, involved the harassment of US naval ships carrying out mapping and surveillance missions in international waters. Meanwhile, Chinese interference with ships from the Philippines, Vietnam, and other Southeast Asian nations raises questions about the freedom of navigation.

Adding to this uncertainty are assertions that the entire South China Sea is Chinese territory, or the controversial nine-dash-line claim, which dates to 1947 and Nationalist China. When the Communist Party of China took over mainland China and formed the People's Republic of China in 1949, it assumed the same territorial policy of the previous government, although it did not press the claim. Since 2010, however, Beijing has turned up the rhetoric and backed it with action. By any account, the claim is unrealistic and is not given credence by its neighbors. But it does emphasize the perspective of China on its sphere of influence in areas it may seek to control.

So far, there has been little disruption to the freedom of navigation through the maritime domain in the Asia-Pacific region. No nation has impeded passage on the high seas or imposed new boundaries. Asian nations usually confine their activities to legally demarcated exclusive economic zones, despite efforts by Beijing to gain acceptance of its expansive maritime claims. However, a presumption exists that innocent passage is not only allowed but defended by maritime nations.[17] The United States and its allies are concerned not only by the threat of the PLAN suddenly trying to rule the waves but also that China will seek to undermine the rule of law in the region. This situation represents a variation of the *broken windows theory* applied to the international community—namely, that interests are weakened by a failure to oppose deviant behavior.[18] The United States has served as the arbiter of regional security through its forward presence and alliances. Should the security break down, US and allied resolve would be questioned. And, should Washington not play the role it has for decades, the dangers of regional instability will increase dramatically.

THE ROLE OF ALLIES

Whether the United States will continue to act as guarantor of regional stability depends to some extent on the strength of the alliance system and congruence of security policies with its partners. Given the tyranny of distance in the Asia-Pacific region, maintaining US presence and credibility relies on its forward-based forces. Moreover, as US interests in the South China Sea and Indian Ocean grow, it will no longer be possible to operate in the region only from Japan; access to ports and airfields in Singapore and other nations will therefore become even more important. Regular sea-lane patrols and air coverage will not be feasible without major weapons platforms and naval and air force infrastructure in the area, particularly in light of the need to respond immediately in the event of instability.

Thus, the resilience of the alliance structure in Asia will significantly determine the ability of the United States to remain a major player. Assessing the strengths of that structure requires attention to the status of alliance relationships among treaty partners.

Japan

The linchpin of US forward presence in Asia continues to be Japan. The home of the Seventh Fleet, Yokosuka Naval Base, and Kadena Air Base, Japan hosts most US military assets in the region. Cooperation between the United States and Japan is especially close on ballistic missile defense and antisubmarine warfare, and both nations also deploy advanced fighter aircraft. Perhaps even more than the United States, Japan has watched the rise of the PLA out of concern that China might threaten the sea-lanes. But like the United States, Japan is wary of antagonizing Beijing and jeopardizing trade. As the Senkaku incident in 2010 revealed, Japan is not immune from Chinese assertiveness in contested maritime areas.

Such concern has shifted the Japanese strategic focus from the northeast and the Russian-held Kurile Islands toward its southwestern islands, or its so-called southwestern wall. A recent decision by Tokyo to purchase F-35 Joint Strike Fighters as its next-generation fighter is a response to advances in Chinese airpower. Similarly, the Japanese plan to expand its submarine force by delaying the retirement of aging boats is driven by increased PLA naval capabilities that include more than sixty submarines.[19] China's military growth has also pushed Japan to seriously engage with other nations to discuss closer cooperation on maritime security.

Greater US-Japan security coordination is hampered by legal restrictions on the Japanese ability to engage in collective self-defense, joint weapons development (though constraints are being relaxed), and overseas deployment. Moreover, bilateral relations have been strained over the relocation of Marine Corps Air Station Futenma to a more remote area of northern Okinawa. The return to power in 2012 of the Liberal Democratic Party foreshadows closer ties after tension with the former ruling Democratic Party of Japan. Yet a continuing desire by the United States for a special relationship with China may limit its embrace of the new Japanese prime minister, Shinzo Abe, and his plans to strengthen Japan Self-Defense Forces.

South Korea

In the last decade the US–South Korean alliance has reached high levels of integration. Stability in South Korea has driven military cooperation and combined training amid swings in policy and domestic antipathy toward the US military. In addition, Seoul has deployed forces in support of almost every US-led military effort since Vietnam to become one of the most reliable American allies. Mutual relations were strained in 2003–2008 when President Roh Moo-hyun advocated a left-wing agenda and closer relations with North Korea. Moreover, anti-American protests are regularly flaring up, such as the incident in 2002 when two schoolgirls were run over and killed by a US armored personnel carrier during a routine training exercise.

Challenges influencing the South Korean relationship with the United States will continue to be associated with domestic politics. The conservative tendencies of President Park Guen-hye seem to be moving Seoul toward closer relations with Beijing and increased tension with Tokyo. Such developments contrast with the policy of former president Lee Myung-bak, who enjoyed close relations with Washington and was willing, at least initially, to explore greater cooperation with Japan. Moreover, the plan to remove the US military presence from central Seoul has been delayed because of environmental and legal issues, thus prolonging this situation as a basis of contention.

A wild card in Seoul-Washington ties is the new leader of North Korea. With the death of long-time ruler Kim Jong-il in late 2011, his youngest son, Kim Jong-un, assumed power. Just two years later, Kim had his uncle and number-two power holder, Jang Song-taek, executed. Early actions indicate that Kim is as aggressive as his father, and his quick decisions on nuclear and ballistic missile tests have returned the Korean Peninsula to a high state of alert. As a result, the United States and South Korea are preparing for continued instability.

Australia

Canberra has long been Washington's closest ally save for the United Kingdom, with Australian soldiers fighting alongside US troops in every overseas operation since World War I. Politically, as well, Australia and the United States have shared common values, which they have promoted in Asia and elsewhere. After 9/11, Prime Minister John Howard was among the staunchest supporters of antiterrorist policy under President George W. Bush. Despite its small military, intelligence and military cooperation by Australia is extremely high. This trend is likely to continue as Australia modernizes and expands its submarine force and purchases highly advanced aircraft, including the US-made F/A-18 Hornet and stealthy F-35 Lightning fighter.

As Washington increasingly focuses on the South China Sea and Indian Ocean, Canberra plays a critical role in maintaining stability in both regions. The announcement in 2011 by President Barack Obama that up to 2,500 Marines would be based in Darwin was the first sign of his so-called pivot to Asia. Protection of sea-lanes to Indonesia from the Indian Ocean can be strengthened by Australia and access to the eastern Indian Ocean is possible from Australia and its territories such as Christmas Island. Like Japan and the United States, Australia is concerned about being encircled or threatened because of its reliance on Chinese markets for minerals and other resources. But the aggressiveness of Beijing in regional waters and its different values have caused Australia to seek greater connection with the liberal nations of the Asia-Pacific region and, particularly, the United States. Moreover, the election of a conservative Australian prime minister, Tony Abbott, in 2013 offered opportunities for greater cooperation on security issues with the United States, given his inclination to support US policies.

Other Allies and Partners

American relations with the Philippines are marked by a colonial past; longtime support of the autocratic leader Ferdinand Marcos, who was popularly ousted in 1986; and the decision not to renew the lease on the US naval base at Subic Bay in 1991. Despite more recent cooperation in training Philippine troops to fight Communist and Islamic rebels, Washington has worked less often with Manila since US bases closed in 1992. Friction with China over the Spratlys and oil reserves in the South China Sea led the Philippines to remind the United States of their mutual defense agreement. Manila also called on Tokyo to assume a larger role in maritime security. Though recent developments have isolated Manila to a certain extent from US policies toward the Asia-Pacific region, it has not decided whether it wants to forge good relationships with Washington and its other allies or tilt closer to Beijing despite regional tensions.

Thailand and Singapore provide key facilities for US air and naval forces, thereby playing important roles in ensuring US forward presence in Southeast Asia. Although neither hosts US forces on a permanent basis, they enable logistical support to maintain US presence. Singapore has expanded this cooperation, sharing information on maritime security and agreeing to offer facilities for the new US Navy littoral combat ship, another plank in the US rebalance policy.[20] Given its location at the intersection of the Malacca Strait and South China Sea, Singapore has strategic importance as a partner. It has taken the lead in organizing regional nations, primarily Indonesia and Malaysia, to combat piracy and promote security cooperation in Southeast Asia, which could address regional concerns on the freedom of navigation.

For its part, Thailand provides US forces access to the Andaman Sea and the Bay of Bengal. In addition, Thailand borders Burma, where China has increasingly sought to increase its presence. However, instability in Thailand in 2006–2010 because of a coup staged against President Thaksin Shinawatra placed bilateral relations on hold as the US regional policy was reexamined. The election of Thaksin's sister, Yingluck Shinawatra, as prime minister in 2011 failed to end domestic unrest, but it did revive high-level contact, including a visit by then secretary of state Hillary Clinton.

India occupies a special role in US security thinking given its size and location. The Indian Ocean is increasingly recognized as a key waterway linking East and West, and increases in commercial and naval traffic have brought new actors, such as China and Japan, into the region with greater regularity.[21] India has announced plans to enlarge its presence in the Andaman and Nicobar Islands, which are situated between the Indian Ocean and Malacca Strait. Concern over Chinese influence in Burma and Pakistan and the possibility of the PLAN gaining access to ports and listening posts has caused India to look for ways to not only increase its capabilities but also cultivate partnerships as a way to quietly hedge against China.

US policy makers talk about a strategic relationship with India that is based in part on democratic values and also on the hope that New Delhi will play a constructive role in maintaining security in the Indian Ocean. New Delhi will be the largest purchaser of naval equipment within the next decade, investing some $50 billion over two decades to buy one hundred ships and submarines.[22] Giving promise to greater movement on regional matters, India, Japan, and the United States held their first trilateral meeting in 2011 to discuss common security concerns.

FUTURE ALLIANCES

Despite concern over China, Asian nations have made it clear that they are not ready to choose sides, unlike Europe during the Cold War. The integrated economies and mutual participation in regional organizations bolster the need for strong ties with China. Additionally, unlike the Soviet Union, Chinese troops today do not occupy any foreign territory except Tibet, although most Asian countries believe Beijing respects national boundaries. What does concern them, however, are contested islands and access to the global commons. Consequently, nations such as India, Indonesia, and Singapore claim that to divide Asia into opposing blocs will force the Chinese to act more aggressively to retain freedom of action. US policy makers are similarly hesitant to endorse a split between Beijing and Washington. They fear that multilateral security alliances might provoke China to become more adversarial, with severe economic consequences, and thus have sought to engage but hedge China while building allied capabilities.[23] Any talk of an Asian NATO is swiftly dismissed as unworkable and unneeded.

The question posed at the beginning of this chapter was whether the half-century hub-and-spoke alliance structure remains the most appropriate vehicle for US strategy in the Asia-Pacific region. Given the concerns of US allies, the current structure seems to have continued relevance. Asian nations long allied with the United States are more interested in maintaining ties in a region that seems to be heading into uncertain times. Yet given the likely challenge to stability, the current structure suffers from a capacity problem and a scope limitation.

In terms of capacity, US allies must develop forces to engage in broader regional initiatives to protect maritime security, freedom of navigation, and national territory from aerial threat. But given budgetary constraints, under which estimated cuts in US military expenditures could reach $1 trillion by the early 2020s, the ability of the United States to maintain a presence in Asia will be severely strained.[24] With older aircraft and dwindling fleets of ships, the Pentagon must begin to decide where to fly and sail, for how long, and under what circumstances. These reductions in funding already have reduced training hours, maintenance, and joint exercises in the Asia-Pacific region. Thus, it is important that US allies such as Japan, South Korea, and Australia expand their militaries to take on a larger role in their neighborhoods.

Collaborating with allies to bolster their capacities is underscored by the mini arms race in Asia. Nearly all nations are buying submarines and many are acquiring advanced fighter aircraft. Surface fleets are growing more slowly, but if the trend continues, increasing numbers of smaller nations will venture beyond their littorals, especially near critical sea-lanes or contested territory. Given contested rules of the road in the maritime domain, there is a need for cooperation among capable naval and air forces to share information and respond to threats to navigation. Moreover, the alliance structure is limited. Although the hub-and-spoke model provides bases that maintain US presence, it is insufficient for the current Asian security agenda.

Northeast and Southeast Asia used to be protected by major bases in Japan and the Philippines, with the United States able to protect its allies and intervene if required to preserve regional stability. After facilities in the Philippines closed, US military posture became unbalanced, with its bases and bulk of forces limited to Northeast Asia. Only port visits and transiting of military units in Asia enabled the United States to maintain its presence in far-flung areas.

Yet as China develops maritime and air capabilities and increases its presence in the East and South China Seas and Indian Ocean, the uncertainty this creates increases the demand for a countervailing US presence. This situation, in turn, will increase demands on US forces, which have operated at high tempo for more than a decade because of the conflicts in Southwest Asia and now face a drawdown in the strength of naval and air capabilities.

US Pacific Command has been considering how to increase access throughout the region. Acquiring new bases is unrealistic in the foreseeable future because of domestic politics and budget constraints in Asia. Plans to base as many as 2,500 US Marines in Australia suggest possibilities of expanding the US military footprint through temporary access in crises. Washington must work with Manila and Bangkok, for example, to facilitate such arrangements. In addition, a far-ranging US strategy must establish partnerships to host forces and pre-position equipment. Logical candidates include Vietnam with its long coastline on the South China Sea, Malaysia with its access to the South China and Andaman Seas, and Indonesia with its national territory stretching from the Indian Ocean to the Western Pacific.

Negotiating access and temporary basing rights with these nations could give US forces ongoing greater reach. It would also create stronger working relations with the three nations in question and bolster efforts to improve their defense capabilities. Moreover, it could begin to form a community with shared interests in the region that is organized around the provision of public goods and preservation of a liberal international order.

There are benefits to protecting distant territories such as Guam and American Samoa, and lesser-known islands such as Johnston Atoll, Midway, the Northern Marianas, and Wake. These areas are isolated in the Pacific Ocean, and US contact with them can be maintained only by the presence of facilities as well as naval and air calls. Guam hosts the most valuable US bases west of Hawaii and east of Japan, although it is not out of reach of either Chinese ballistic missiles or the PLAN in the Pacific.[25] Defense planners are worried about losing "white space" to China—that is, losing influence over unaligned, usually smaller nations because of China's operational advantage over areas closer to the mainland than the continental United States where Beijing is willing to invest development aid to secure trade concessions and landing points. Recently the Chinese have focused on Fiji, Kiribati, and Papua New Guinea. As US access is extended, China may respond by concentrating assets closer to home. Furthermore, as allies and partners increase their sea-lane patrols and joint response to regional instability, the US Navy and Air Force can pay greater attention to beefing up defenses on Guam or making port visits to Samoa and other territories. Such a policy will help maintain a balance of power in Oceania that can blunt the Chinese effort to extend its influence.

What kind of structure does the United States need to augment the hub-and-spoke system? The system, along with its obligations, should remain central to US policy toward Asia. However, Washington should take the lead by creating a new community of interests that includes both old allies and new partners. Future cooperation will be facilitated with nations such as Vietnam with its communist regime if it is conducted by partners that have varying relationships with the United States. One approach would be to develop a concentric geographic structure with outer and inner relationships. The outer would couple Australia, India, Japan, and South Korea: three allies and a nation with which the United States hopes to build a strategic partnership. The inner would connect Indonesia, Malaysia, the Philippines, Singapore, Thailand, and Vietnam to anchor security cooperation in the Indo-Pacific region and US foreign policy. Each of the nations has greater relative capacity than other nations in the Asia-Pacific region and also has served as a leader in its respective subregion. The relationships among the inner group would enhance security in the littorals and guarantee access to the South China Sea.

With initiatives such as antipiracy cooperation among Indonesia, Malaysia, and Singapore, the navies and air forces of the inner group could be trained and supported by the outer group and the United States. Although this arrangement may influence nations of the outer group to exercise more responsible security roles, the nations of the inner group could be encouraged to adopt more liberal policies by cooperating with more democratic nations.

• • •

A concentric strategy is not designed to contain China; rather, it is a flexible instrument to bolster the long-standing US alliance system in Asia. As the region becomes more integrated in a larger Indo-Pacific community, promoting both international standards of conduct and support for the liberal international order is more important than ever. Such a two-tiered US approach to traditional allies and new partners offers a vision of how peace and prosperity can be maintained throughout Asia and will contribute significantly to promoting the common goals and protecting the interests of the United States and all the nations of the Indo-Pacific region.

NOTES

1. Ronald Spector, *In the Ruins of Empire: The Japanese Surrender and the Battle for Postwar Asia* (New York: Random House, 2008).

2. James Mann, *About Face: A History of America's Curious Relationship with China from Nixon to Clinton* (New York: Vintage Books, 2000).

3. "Wen Warns US on South China Sea Dispute," BBC News, November 18, 2011, http://www.bbc.com/news/world-asia-15790287.

4. Steven Hildreth, "North Korean Ballistic Missile Threat to the United States" (Washington, DC: Congressional Research Service, Library of Congress, 2009), 3.

5. Steve Herman, "North Korea Reported to Have Test-Fired Anti-ship Missile," *Voice of America*, November 15, 2011, http://www.voanews.com/content/north-korea-reported-to-have-test-fired-anti-ship-missile-launched-from-bomber-133949028/148244.html.

6. International Institute for Strategic Studies, *The Military Balance 2011* (London: Routledge, 2011).

7. Robert Johnson, "India Is Preparing to Sell BrahMos Supersonic Cruise Missiles to Vietnam," *Business Insider*, September 20, 2011.

8. A press release issued by the Japan External Trade Organization noted record-breaking growth compared to 2010. Japan External Trade Organization, "Japan-China Trade Sets Record in First Half of 2011, While Earthquake Slows Growth in Exports," July 23, 2011.

9. Jeremy Page, "China Suspends Talks with Japan," *Wall Street Journal*, September 19, 2010; Kyung Lah, "China Arrests 4 Japanese Against Backdrop of Diplomatic Battle," CNN, September 24, 2010, http://www.cnn.com/2010/WORLD/asiapcf/09/24/china.japanese.arrests; Keith Bradsher, "China Restarts Rare Earth Shipments to Japan," *New York Times*, November 19, 2010.

10. Ben Bland and Kathrin Hille, "Vietnam and China Oil Clashes Intensify," *Financial Times*, May 29, 2011.

11. Keith Bradsher, "Philippine Leader Sounds Alarm on China," *New York Times*, February 4, 2014, http://www.nytimes.com/2014/02/05/world/asia/philippine-leader-urges-international-help-in-resisting-chinas-sea-claims.html.

12. "S. Korea Hits Japan in Escalating Territorial Dispute," *USA Today*, March 18, 2005.

13. Amir Tejo, "Navy Was Set to Fire on Warship," *Jakarta Globe*, June 4, 2009.

14. "Clinton: Senkakus Subject to Security Pact," *Japan Times*, September 25, 2010.

15. Keith Richburg, William Wan, and William Branigin, "China Warns US in Island Dispute," *Washington Post*, June 23, 2011.

16. Sheldon W. Simon, "Safety and Security in the Malacca Strait: The Limits of Collaboration," in *Maritime Security in Southeast Asia: US, Japanese, Regional, and Industry Strategies* (Seattle: National Bureau of Asian Research, 2010), 3.

17. For a more general discussion, see James Kraska, *Maritime Power and the Law of the Sea: Expeditionary Operations in World Politics* (New York: Oxford University Press, 2011).

18. James Q. Wilson and George L. Kelling, "Broken Windows: The Police and Neighborhood Safety," *Atlantic Monthly*, March 1982, pp. 29–38.

19. Eric Talmadge, "Japan, Worried by China, May Boost Submarine Fleet," *Washington Times*, October 26, 2010, http://www.washingtontimes.com/news/2010/oct/26/japan-worried-china-may-boost-submarine-fleet/?page=all.

20. Shaun Tandon, "US Navy Expects to Base Ships in Singapore," Agence France Presse, December 17, 2011.

21. Robert Kaplan, *Monsoon: The Indian Ocean and the Future of American Power* (New York: Random House, 2011).

22. "India Eyes Mega-Navy Spend $50 bn," *Indian Express*, May 16, 2011.

23. Aaron Friedberg, *A Contest for Supremacy: China, America, and the Struggle for Mastery in Asia* (New York: W. W. Norton, 2011), 1–9.

24. "Obama's Defense Drawdown," *Wall Street Journal*, January 9, 2012.

25. Paul Barton, "Strategist: Guam in Reach of Chinese Missiles," *Air Force Times*, March 30, 2011.

6 STRATEGY AND CULTURE

Colin S. Gray

STRATEGIC CULTURE would be denied a certificate of airworthiness if it were a plane. However, it does fly and can take us to places where knowledge might be found. The concept is flawed in terms of both logic and evidence, but such weaknesses should not be considered fatal. Scholars who question its legitimacy, uttering, "So what?" should do so objectively when debating strategy and culture. Many people fail to acknowledge that discovering the inherent weaknesses of an idea does not discount or cheapen the value of discussing it.

This chapter argues that cultural perspectives on strategy are seriously flawed but helpful and essential. The argument would equate to defending the indefensible only if the merit found in such perspectives suppressed treating cultural analyses with necessary reservations firmly in mind. If one asks too much of cultural perspectives, they can become dangerously misleading. One must not forget that strategy—grand and military—involves life and death writ large; ideas matter because of their potential impact on people. Thinking strategically is a practical endeavor, which means that attitudes, beliefs, and values with cultural import should not be gauged in terms of logic alone. Inferior ideas on strategy are not simply unpalatable; they can lead to death and destruction and also to political and strategic consequences.

Sun Tzu was correct in asserting the significance of self-knowledge and understanding the enemy but less accurate in claiming that such insights guarantee victory.[1] Both knowledge and understanding enable strategic effectiveness, yet in and of themselves they neither sink ships nor win hearts and minds. While knowing oneself and one's enemy may be appealing, it is not sufficient to achieve victory. History has shown that enemies have been successful despite being culturally ignorant. Moreover, enemies have prevailed without an explicit and consistent strategy worthy of the name. Cultural ignorance and contempt for strategy can have lethal consequences, but excellence on the tactical and operational levels may offer sufficient net strategic effectiveness to meet political requirements.[2]

Competing analyses have discovered more and less in investigating cultural influences on strategy. If strategies must be great or small in cultural terms, then the question of whether strategic culture is legitimate remains relevant.[3] Strategies also must encompass the human, political, geographic, and technological. One unfounded argument goes as follows: all strategies are devised and executed by encultured individuals, thus all strategies are influenced by culture. Insofar as enculturation is unavoidable, strategic actions always bear some cultural stamp, if ever so faint. Aside from being trivial, this argument misleads because it marginalizes strategic actions. One must take account that strategic history results from competition. Given that the strategy requires engaging with others, how valid are claims that it is culturally shaped, and if so, how much? What happens when strategic cultures meet? And if protagonists have disparate cultures, the apposite cultural landscape may be too complex to be analyzed with any degree of confidence.

The influence of culture simply asserts the identity of actors or organizations. Even when the United States must respond to strategic circumstances shaped by a foreign strategic culture, that response will bear some American characteristics. However, it is not self-evident that those characteristics will be particularly American. To state the matter bluntly, the notion of strategic culture is logically and empirically problematic. One cannot presume identity drives behavior. More often than not, identity is overcome by exigency. People tend to behave in ways not only because of who they are and what they believe in but also because of where they find themselves, and not necessarily by their own volition, either politically or strategically.

CONCEPTS AND DEBATE

The concepts of culture and strategy are individually contestable, but their merger can produce devilish confusion and mischief. Talking about strategy and culture allows useful analytical discretion, while the familiar concept of strategic culture invites the act of reversal to read seductively as cultural strategy. This syntactical sleight of hand is instructive because it suggests that strategy must be cultural. Indeed, could there be any strategy that is not cultural?

Definitions are arbitrary but vary in quality of purpose and authority. It is also necessary to remember that social science theory is understood as providing most-case explanations. Some theory can have predictive value, but the population size of the relevant event must be large and cannot deliver reliable predictions on individual happenings. Those seeking correct answers to quintessentially strategic questions by quantitative analysis are analogously pursuing objective truth in astrology through better methodology. Such attractive terms as a calculus of deterrence or strategic arithmetic may flow from writers' pens, but they are fool's gold. Indubitably, strategies intended to deter involve some calculation, but the most vital currency conversions—from military to strategic effect and from strategic to political effect—do not lend themselves to being calculated and therefore to being verified mathematically.

The terms *strategy* and *strategic* are often misused, but the following definition has some authority: the direction and use made of force, and the threat of force, for purposes of policy as decided by politics.[4] As for defining *culture*, we are spoiled for choice. With culture and strategic culture one enters an unregulated realm. Given that *strategy* and *strategic* are used haphazardly, *strategic culture* is unlikely to unlock the door to knowledge precisely. Although some researchers have confidence in strategic culture enquiries, only true believers in forensic qualities put trust in discovering something akin to scholarly keyhole surgery. In fact, culture tends to be more problematic as a conceptual tool because it is scrutinized for its postulated strategic type. Nonetheless, the concept of strategic culture does survive more or less intact, albeit battered, as significant for an understanding of strategy.

There is no correct definition of strategic culture, and it would be an exaggeration to claim that one is more authoritative than another.[5] However, the debatable concept of strategic culture relating to the assumptions, beliefs, attitudes, habits of mind, and preferred modes of behavior, customary behavior even, bearing on the use of force by a security community has been adopted. The core meaning of culture is signified by the notion of a common stock of cultural references shared by members of a community, but what makes this common stock cultural?[6]

The literature on strategy and culture is impaired by a hunt for professional-looking social science. It has resulted in the paradoxical and ironic consequence of hindering what can be done in the interest of acquiring the impossible. Good practice in developing theories is challenged by the fundamental nature of strategic culture. The question is where does culture end and influence begin? The solution lies in substantiating that culture is strictly ideational.[7] Concepts that appear to qualify as cultural because of their persistence and popularity then may be assessed for evidence that give them expression. Strategy is cultural only to a degree because it is more than culture by definition.[8] The cultural analysis of strategic phenomena should not be seduced by the belief that some theory exists with predictive value for policy and merely is awaiting discovery. Instead, the theory connecting strategy and culture offers insights that cue recognition of patterns in thought but is valuable only within the context of a general theory of strategy.

Definitions of culture abound, and those of strategic culture offer choices that are certain to enable scholars of most persuasions to find one they like. Lawrence Sondhaus has performed a signal service by providing a careful, clear yet nuanced analysis of the definitional choices. The question that remains in dispute is the nontrivial one of identifying the subject. Does one seek the influence of culture only on strategic phenomena, or is culture found on and in those phenomena? Scholars agree that culture is ideational, but some believe it can be identified in material objects as well as behavior. Thus subject and object merge to challenge methodological analysis, which is admittedly controversial. Defining strategic culture insults the minimalist doctrine of William of Ockham. Fear of omissions has been harmful through needless specificity.[9]

One peril of analyzing strategic culture is distinguishing independent variables, which can be severe when making allowances for probable consequences of feedback. Because encultured people execute strategy universally, how can the influence of culture be isolated? Even dodging this problem by insisting that strategic culture must be tested is conceptual. There is no way of dividing strategic ideas and practices.[10] Strategic behavior is influenced by a conceptual culture that is shaped from past and current strategic behavior.

Understood strictly in ideational terms, strategic culture cannot be regarded as a conceptual variable effectively independent of strategic practice. Strategic ideas emerge to a greater or lesser extent from what is learned from both practice and malpractice, while strategic practice is strategic theory in action in the field no matter how imperfectly it is applied. And when long-held strategic ideas are refuted rather than validated by contemporary strategic experience, conceptual and other culture can and often does change.

As with the concept of strategy, which has at the core of its meaning the connections between ends, ways, and means and is focused on coherent instrumentality and consequences, so strategic culture can be understood as having a core meaning with some intellectual integrity. Culture is important to the understanding of strategies because it directs attention to the customs, beliefs, and behaviors that are persistent and therefore presumably relatively deep rather than ephemeral and shallow. To identify a decision as being strategically cultural might suggest that it reveals not merely a passing opinion but rather an attitude expressing enduring assumptions and beliefs. A noteworthy contributor to the challenges in exploring strategy and culture, let alone the two together, is that each is lacking an analytically convenient boundary to its domain. Just about anything may contribute some strategic effect, while culture also is really frontier porous if not free. Elsewhere, it has been argued that strategic culture provides a context for strategic decisions and action, but Antulio Echevarria is correct in claiming that context in itself has no objective end.[11] The problem lies not with the intellectual integrity of the concept but rather with all contexts having a context without any empirical limits.

The debate over strategy and culture is familiar to historians of strategic ideas. Accepting this is important because it provides a much-needed perspective. When seen in the light of military, defense, and strategic debates, controversy over culture becomes easier to understand. Intellectual historians are bent on discovering new schools of thought, which form rival camps. The great strategic debates since 1945 have recorded attack, counterattack, and an outcome, if not conclusion, that is synthetic. For example, after a brief epiphany in 1946, the debate over nuclear weapons moved from imprudent enthusiasm to no-less-imprudent reluctant acceptance.[12] The recent interest in cyberpower and cyberspace strategy signals anxiety, with a thesis of maximum cyberpotency in ascent.[13] Inevitably, second thoughts will trigger antithetical claims that cyberspace has become another menacing global domain.

The debate over counterinsurgency (COIN) and counterterrorism (CT) has shifted in favor of cultural influences on strategic studies.[14] COIN all but vanished from the agenda of strategic studies in the wake of the Vietnam War. The defense community met the challenges of Afghanistan and Iraq with its usual enthusiasm for the fashionable and faddish. COIN and CT were rediscovered to the self-satisfaction of many, if only in cultural terms.[15] This development responded to the functional needs for COIN doctrine, training, and equipment in the last decade. For COIN, which normally involves war among people and unarguably is always war about people, one key to success has been cultural appreciation of societies in contention.[16] This cultural turn in strategic studies has resulted from the undeniable political and military failures in Southwest Asia.

The US defense community remained uninterested in the study of strategic culture for decades. The so-what attitude was not the skeptical retort of questioning strategic minds but contemptuous dismissal. While defying the odds, strategic culture achieved limited purchase in the realm of nuclear strategy in the 1970s and slipped under the radar of serious official attention in the 1980s and 1990s.[17] Why did the only superpower need to study the culture of potential adversaries? The rediscovery of cultural appreciation as a strategic enabler has been overcelebrated. For prophet-advocates, cultural understanding could make the vital difference between success and failure. If war is cultural behavior as some claimed, surely, the argument went, cultural appreciation was bound to yield strategic rewards.[18]

Predictably, the cultural turn in US strategic thinking, and less so in strategic practice, was overadvertised as a magic bullet. In addition, the emphasis on cultural understanding triggered counterrevolutionary reactions, which were general and specific. The assumption of culturalism as applied to strategy was challenged, as were the benefits attributed to COIN and CT doctrines. Analysts have critiqued strategic culturalism and the cultural tilt in COIN.[19] The strategy and culture debate has moved from being largely a bone of academic contention to official acceptance and popular acclaim to the point that it has been encountering purposeful blowback and the idea of culturalism has been seriously challenged.[20]

This debate promotes exaggeration and theorizing demands reductionism in presenting, if not treating, complexity. Both cultural and countercultural arguments have been oversimplified. Given its past neglect in strategic studies, it is tempting to excuse some overenthusiasm for the cultural perspective.[21] Nonetheless, balance is required for a reliable understanding of strategy. The pertinent challenge is identifying the elements of the cultural narrative that can withstand scrutiny of particular arguments as well as general criticism.

PROTOTHEORY

In the words of Carl von Clausewitz, "Theory should cast a steady light on all phenomena so that we can more easily recognize and eliminate the weeds that always spring from ignorance."[22] In the course of the last four decades, numerous studies have sought to unravel the mysteries of the relationship between strategy and culture.[23] There has been much assertion and argument, some of which has been useful for scholars. But while the ground has been cleared and rival positions staked out, the time has come to formulate a general theory. Both thesis and antithesis benefit as well as impair the pursuit of this endeavor. It is necessary to review the intellectual landscape of the debate over strategic culture to ascertain what of value remains in the wake of conceptual battles that warrants being maintained for further development.[24]

It may be helpful to register though not necessarily endorse the view that culture is a high concept that invites the nominalist fallacy. If one simply discarded the concept of culture and let it pass out of intellectual fashion and common usage, nothing of forensic value would be lost for strategic inquiry, so the argument proceeds. One need not endorse this proposition to examine culture as having value for study and, as Echevarria suggests, strategic practice.[25] There is some merit in criticisms of the cultural argument and purposefully culturally flavored analysis of strategic behavior. Unsurprisingly, the relatively easy mission is the staking out of cultural argument, on the one hand, and discovering those arguments to be flawed, on the other. The legal course is best pursued through an adversarial method. Thus far the scholarly debates have identified positions, registered claims, and created intellectual vistas with riotous displays of weeds and flowers—not that the two always can be classified accurately with any confidence. That granted, the literature on strategic culture (or culture and strategy) has yielded candidates for inclusion in a general theory to advance understanding of this subject.

Culture exercises only a conditioning influence, and thus it cannot be operationalized like predictive analytical tools. This is not to imply that culture has little or no practical utility but that the utility is constrained by the lasting though highly variable nature of strategy. In common with grand concepts, culture and even strategic culture are misleadingly imperial on inspection. It is the nature of scholarly inquiry for research to claim evidence that makes ever more precise classifications. A much-heralded revolution in military affairs (RMA) that excited US and other analysts in the 1990s required being augmented by some larger concept and turning to the origins of Soviet theories on military-technical revolutions. Grammatical deference to RMA exploited the term and also provided empirical reality. The RMA concept was theory.[26] Likewise, some researchers took it as strategic culture, but time and forensic evidence revealed that its conceptual species had relatives—namely, public (and possibly) civic culture and military culture. It became convenient to group these subjects under the rubric of strategic culture but in common with the national way of war and style in strategy, concepts potentially damaged by being underevidenced and relying on doubtful assumptions.[27]

The relations between what should be subcultures within a supposedly simple national cultural domain and between subcultures and grander national-level cultures (public or strategic), remain undergoverned intellectual space. How does an impressive conceptual postulate based inter alia on American strategic culture, way of war, and national strategic style hold up when subjected to theories explaining geographic and functional military cultures? This is not to claim that service or military cultures harm the unitary national strategic culture, but it suggests the possibility. The evidence supports what might be called subsystem strength. Some works point to the need for discriminating and holistic strategic and military cultures.[28] The development of theory would be challenged if it were known that soldiers, sailors, airmen, space warriors, and cybernauts had more in common culturally with people who are geographically and functionally in the same domains abroad than with their sister services. Similarly, some might find it troubling to discover that the special operations community has been politically frontier-free when it comes to culture. Is it possible that Chinese cyberwarriors fight in a distinctively Chinese way?

Tactical choices are influenced and sometimes determined by particular geographic and historical contexts of national militaries. It is not difficult to detect the cultural stamp on tactics and weaponry. However, to what extent do pilots, submariners, or artillerymen share a common worldview because of branch specialties? Can aircraft be flown in a Chinese style? One answer may be that while military hardware requires some common technical skills, the tactics to apply those skills could differ markedly. Even when military cultures and subcultures have national or local hallmarks, human behavior owes more to specific instruments than to cultural authority.[29] Dominick Graham expressed this idea neatly in analyzing the performance of the British Army in two world wars in noting that "each arm of the army had a distinct view of the battlefield."[30] The implications of this claim tend not to attract the notice they deserve.

When one begins to consider seriously the proposition that soldiers, sailors, airmen, and others may have worldviews sufficiently distinctive as arguably to merit description as cultural, insight should be gained on some of the endemic problems in joint and combined warfare. Much intra- and interservice disharmony is rooted in culture. Each geographic focus has fostered strategic attitudes that can prove severely dysfunctional in efforts that need to be joint and combined for strategic purposes. Understandably, armies tend to believe that warfare is about controlling land and people through presence and occupation; navies necessarily approach warfare as a struggle for lines of communication, and air forces are likely to regard warfare as an exercise in targeting for kinetic effect from altitude, viewing the strategically relevant world as akin to a dartboard. These cultural leanings, albeit oversimplified for emphasis, are complementary and can gel to make them greater than the sum of their parts, which often is not the case. Instead of coherent joint and combined military efforts, reality frequently is the conduct of many styles in warfare, loosely lashed together, and sponsored rather than commanded and controlled by a nominally unified command structure. The point here is that culture has value for understanding diversity and its challenges with respect to what needs to be joint and combined.

The dynamism of statecraft and strategy in peace and war means cultural information must be valued for its educational worth. The general theory of strategy involves education, though it is not a "sort of *manual* for action," as Clausewitz noted.[31] Cultural understanding of enemies is not an adequate basis for strategic decision. What such understanding achieves, even assuming it is superior, is comprehension of enemy values and probably inferable preferences, which may be important but does not include the full measure of considerations, both enduring and ephemeral, that result in decisions by an enemy at particular times and places.

Culture must have enduring qualities to be more than opinion. Much as historical research on evidence for the RMA tends to locate revolutions in abundance, so the cultural postulate can hardly avoid biasing analysts in favor of persistence over change, custom over innovation. This can be unfortunate because it is in the nature of culture to change radically and suddenly, though more often slowly by evolution. It takes time to embed culture and custom. If the cultural thesis has merit, then it should incorporate an ability to accommodate change by plausibly arguing that there is an RMA with Chinese characteristics.[32]

Openness to innovation, especially when it is technological, is thoroughly compatible with a cultural leaning.[33] Nonetheless, scholars must be alert to rapid cultural change through learning and borrowing when circumstances provide the motivation. Regardless of how strategic culture is defined, it cannot become a matter of current intellectual discretion. Only the passage of time reveals whether decisions merit being classified as cultural. Americans may leave America but are unlikely to cease being seen as American, especially to themselves.

The cultures people acquire have not been selected from a catalog of alternative culture options. Particular cultural actions are preferred because they acquire moral authority by being seen as right instead of wrong according to local ethical codes. And the moral authority of the pertinent ethical code will rest on an understanding of the historical narrative of grand strategy. Regardless of whether the culturally authoritative narrative is entirely mythical or substantially demonstrated, culture is about belief, which is to say subjective not objective truth. Self-styled problem-solving people are mistaken in neglecting the influence of culture. Mind usually rules over muscle.[34] It is critical to remember that there is some essential circularity in cultural influence. Cultural themes are thought to bear on strategic success, though remembrance of heroic failure can have cultural significance. Consider the Serbian defeat in Kosovo in 1389, the Lost Cause that shrouds the Confederacy, and the British miracle at Dunkirk.

ESSENTIAL BUT PROBLEMATIC

Scholarly combat has been waged, episodically but vigorously, and some initial arguments can be classified as sick though breathing while one or two are dead if not yet buried. Culture reigns universally over strategy, because all strategy is devised and executed by encultured people. But whereas culture must reign, it does not always rule. Unless one is careless with one's definition of culture and places no defensible boundary around the cultural domain, only rarely is cultural preference afforded sovereign command authority over a community's strategy.

It is ironic and even paradoxical to note that while it is easier to demolish than defend the cultural interpretation of strategic behavior at least as a relevant factor, usually it contributes to plausible explanations. Methodologically, one might grant cultural argument a passing grade of the kind that really translates as a condoned fail. But cultural analysis generates insight despite weaknesses in the methodology on which it rests uneasily. For strategic scholars and operators, the half-full glass of culture has unreliable value. It is an error to discount cultural explanations of strategic history because they do not pass rigorous examination by the social sciences. Most historians will not lose sleep over this judgment. The fact remains that culture speaks volumes to the moral and political vision underlying the politics that generate policy to guide particular strategic thought. Cultural influence is potentially important, but it is not always, let alone determinatively, relevant to specific strategic decisions.

In deference to the familiar notion of strategic culture, which is a combination of big ideas compounded by brutal reductionism as an overly simple unified concept, the subject is strategy and culture. Granularity is a feature of scholarship. The more academics examine phenomena, the more they resemble Victorian entomologists, finding an ever-greater number of distinctive beetles or, in this instance, strategic cultures and subcultures. Typically, military establishments contain multiple cultures, though such potential confusion can be improved by acknowledging loyalty to queen, country, regiment, in whatever order preferred. Nevertheless, the concept of national strategic culture has sufficient merit to act as a source of insight. This does not imply that insight serves as a reliable tool. The culminating point of victory for a strategist in cultural analysis occurs swiftly and over less challenging empirical terrain.

A logic and tactical and operational grammar for each level confronts the process of command and control for well-balanced strategic performance.[35] Even if one uncovers what appear to be cultural phenomena in operational artistry or tactical behavior, the phenomena may be superficial when compared to influences that are transcultural. Ideas on best practices, or doctrine, vary among militaries especially in peacetime, when they are bereft of the discipline of firsthand experience. But actual combat experience serves as a great educator and trainer, just as cultural preference is tested and often revised, sometimes radically.

A problematic claim that values and interests converged in the case of Kosovo was made by Prime Minister Tony Blair in a speech to the Economic Club of Chicago in April 1999 that became known as the Blair Doctrine. Ironically in its fallaciousness, his declaration has served useful purposes. For example, because American values can be summarized in the triptych of liberty, democracy, and the free market and given that they are primarily cultural, it is unarguable that culture often does not control statecraft and strategy. Considerations of Realpolitik usually are not overruled by some policy requirement, which is obviously cultural in the narrow sense of being value dominated. Cultural emphasis cannot be regarded as the determinative factor in instigating political decisions and the consequent strategic action.

• • •

National style and culture as influences on strategic behavior and national way of war and peace should be retained as part of the scholarly apparatus because of their potential value for understanding and explanation.[36] Such expansive ideas should be handled with care, because when misapplied beyond their analytical reach they can be perilously unsound. Unhappily for scholarship, notwithstanding its general endorsement by John Keegan, the proposition that there remain contrasting and apparently all but eternal Eastern and Western ways of war have been demonstrated to be improbable oversimplifications of sharp alternatives, a fallacy traced to Herodotus.[37] Some conceptual formulae are too imperial to withstand severe empirical challenges. When exceptions exceed the norm, it is time to waive the rule. Regrettably, intellectual boldness is not a sufficient test for the merits of a theory.

The perceived value of the cultural perspective on strategy is maltreated by hyperbole and uncritical acceptance, but the quality of the insights that it provides is beyond serious challenge. The cultural glass is half full and cracked, but it offers an essential and unique understanding. It is nonetheless true that cultural insight cannot forecast strategic behavior, which is too complex to yield to the monocausal powers of analytical assault and diagnosis.

NOTES

1. Sun Tzu, *The Art of War*, trans. Samuel B. Griffith (Oxford: Clarendon Press, 1963), 84.

2. On strategic effect and effectiveness, see Colin S. Gray, *The Strategy Bridge: Theory for Practice* (Oxford: Oxford University Press, 2010), chap. 5.

3. Ibid., 16

4. My definition is Clausewitzian. See Carl von Clausewitz, *On War*, trans. Michael Howard and Peter Paret (Princeton, NJ: Princeton University Press, 1976).

5. A range of definitions is examined in Lawrence Sondhaus, *Strategic Culture and Ways of War* (London: Routledge, 2006), 124–25.

6. See "Honest to God," *Times* (London), April 22, 2011, which explains the cultural significance of the King James Bible of 1611.

7. Alastair Iain Johnston states that his book "is about ideas and their relationship to behavior." *Cultural Realism: Strategic Culture and Grand Strategy in Chinese History* (Princeton, NJ: Princeton University Press, 1995), ix. He defines *strategic culture* as "an

integrated system of symbols (e.g., argumentation structures, languages, analogies, metaphors) which acts to establish pervasive and long-lasting strategic preferences by formulating concepts of the role and efficacy of military force in interstate political affairs, and by clothing these conceptions in such an aura of factuality that the strategic preferences seem uniquely realistic." Alastair Iain Johnston, "Thinking About Strategic Culture," *International Security* 19 (Spring 1995): 46. My definition has oscillated about an earlier version, to wit: strategic culture comprises "persisting socially transmitted ideas, attitudes, traditions, habits of mind, and preferred methods which are more or less specific to a particular security community that has had a unique historical experience." Colin S. Gray, *Modern Strategy* (Oxford: Oxford University Press, 1999), 131. These definitions illustrate different approaches in Sondhaus, *Strategic Culture and Ways of War*, 124. Others are provided by David G. Hagland, "What Good Is Strategic Culture?" in *Strategic Culture and Weapons of Mass Destruction: Culturally Based Insights into Comparative National Security Policymaking*, ed. Jeannie L. Johnson, Kerry M. Kartchner, and Jeffrey A. Larsen (New York: Palgrave Macmillan, 2009), 16–19, and Thomas G. Mahnken, "U.S. Strategic and Organizational Subcultures," in Johnson, Kartchner, and Larsen, *Strategic Culture and Weapons of Mass Destruction*, 70.

8. See Gray, *Strategy Bridge*, 59–60.

9. Clausewitz provides a concept that should be adopted as a guide to sufficiency—namely, "the culminating point of victory," which can be stated as more is less. *On War*, 566. The point can be made with fewer words than required by some (especially official) publications. One work that illustrates encyclopedism or vocabulary creep is the *Department of Defense Dictionary of Military and Associated Terms*.

10. John A. Lynn, *Battle: A History of Combat and Culture* (Boulder, CO: Westview, 2003), focuses heavily on the conceptual dimension to cultural influence on strategy.

11. Gray, *Modern Strategy*, chap. 5; Antulio J. Echevarria II, "American Strategic Culture: Problems and Prospects," in *The Changing Character of War*, ed. Hew Strachan and Sibylle Scheipers (Oxford: Oxford University Press, 2011), 432.

12. See Bernard Brodie, *The Absolute Weapon: Atomic Power and World Order* (New York: Harcourt, Brace, 1946); Colin S. Gray, *Strategic Studies and Public Policy: The American Experience* (Lexington: University Press of Kentucky, 1982); Lawrence Freedman, *The Evolution of Nuclear Strategy*, 3rd ed. (Basingstoke, UK: Palgrave Macmillan, 2003); Keith B. Payne, *The Great American Gamble: The Theory and Practice of Deterrence from Cold War to the Twenty-First Century* (Fairfax, VA: National Institute Press, 2008); and Beatrice Heuser, *The Evolution of Strategy: Thinking War from Antiquity to the Present* (Cambridge: Cambridge University Press, 2010), chap. 14.

13. Franklin D. Kramer, Stuart H. Starr, and Larry K. Wentz, eds., *Cyberpower and National Security* (Washington, DC: National Defense University Press, 2009);

Richard A. Clarke and Robert K. Knake, *Cyber War: The Next Threat to National Security and What to Do About It* (New York: HarperCollins, 2010); David J. Betz and Tim Stevens, *Cyberspace and the State: Toward a Strategy for Cyber-power* (London: Routledge, 2011). For two vigorous arguments that frame the debate, see John Arquilla and David Ronfeldt, "Cyberwar Is Coming," in *In Athena's Camp: Preparing for Conflict in the Information Age*, ed. John Arquilla and David Ronfeldt (Santa Monica, CA: RAND, 1997), 23–60; and David Beitz, "Cyberwar Is Not Coming," *Infinity Journal* 3 (Summer 2011): 21–24.

14. Patrick Porter, "Good Anthropology, Bad History: The Cultural Turn in Studying War," *Parameters* 37 (Summer 2007): 45–58; Montgomery McFate, "Culture," in *Understanding Insurgency: Doctrine, Operations, and Challenges*, ed. Thomas Rid and Thomas A. Keaney (London: Routledge, 2010), 189–204.

15. *The US Army/Marine Corps Counterinsurgency Field Manual* (Chicago: University of Chicago Press, 2007); David Kilcullen, *The Accidental Guerrilla: Fighting Small Wars in the Midst of a Big One* (London: C. Hurst, 2009). Daniel Marston and Carter Malkesian, eds., *Counterinsurgency in Modern Warfare* (Botley, UK: Osprey, 2008), provides historical perspectives but lacks detachment from what can be regarded as a counterinsurgency project. Undue attention to culturalism is assaulted in Rob Johnson, *The Afghan Way of War: Culture and Pragmatism: A Critical History* (London: C. Hurst, 2011).

16. Rupert Smith, *The Utility of Force: The Art of War in the Modern World* (London: Allen Lane, 2005), xiii.

17. On this era see Jack L. Snyder, *The Soviet Strategic Culture: Implications for Limited Nuclear Operations* (Santa Monica, CA: RAND, September 1977); and Colin S. Gray, *Nuclear Strategy and National Style* (Lanham, MD: Hamilton Press, 1986).

18. John Keegan, *A History of Warfare* (London: Hutchinson, 1993).

19. Patrick Porter, *Military Orientalism: Eastern War Through Western Eyes* (London: Hurst, 2009); Gian P. Gentile, "A Strategy of Tactics: Population-centric COIN and the Army," *Parameters* 39 (Autumn 2009): 5–17; Johnson, *Afghan Way of War*. An earlier assault was mounted in Michael C. Desch, "Culture Clash: Assessing the Importance of Ideas in Security Studies," *International Security* 23 (Summer 1998): 141–70.

20. Alastair Iain Johnston, *Cultural Realism: Strategic Culture and Grand Strategy in Chinese History* (Ithaca, NY: Cornell University Press, 1995), 4–22; this major study critiqued works with cultural leanings, including my own, to which I replied in *Modern Strategy*, chap. 5.

21. A book that has stood the test of time is Ken Booth, *Strategy and Ethnocentrism* (London: Croom Helm, 1979), which reads today as well as it did when it was written, possibly better.

22. Clausewitz, *On War*, 578.

23. Notable works include Johnston, *Cultural Realism*; Peter J. Katzenstein, ed., *The Culture of National Security: Norms and Identity in World Politics* (New York: Columbia University Press, 1996); Lynn, *Battle*; Sondhaus, *Strategic Culture*; Johnson, Kartchner, and Larsen, *Strategic Culture*; Porter, *Military Orientalism*; and Johnson, *Afghan Way of War*.

24. Echevarria claims that in "policymaking, where some degree of uncertainty is unavoidable, studies of strategic culture do not appear to reduce it . . . [but] help indirectly by offering conclusions that can be challenged and critically examined." "American Strategic Culture," 442. His argument recalls the rationale of the Anglo-American Combined Bomber Offensive—namely, that by forcing the Luftwaffe to fly at certain altitudes it could be destroyed through attrition. Echevarria may be correct, but I find that a more positive view of the cultural perspective with a healthy pinch of salt is advisable.

25. Ibid.

26. MacGregor Knox and Williamson Murray, eds., *The Dynamics of Military Revolution, 1300–2050* (Cambridge: Cambridge University Press, 2001); Dima Adamsky, *The Culture of Military Innovation: The Impact of Cultural Factors on the Revolution in Military Affairs in Russia, the US, and Israel* (Stanford, CA: Stanford University Press, 2010).

27. Gray, *Nuclear Strategy*, was an early undernuanced venture.

28. J. C. Wylie, *Military Power: A General Theory of Power Control* (Annapolis, MD: Naval Institute Press, 1989); Carl H. Builder, *The Masks of War: American Military Styles in Strategy and Analyses* (Baltimore, MD: Johns Hopkins University Press, 1989); Brian McAllister Linn, *The Echo of Battle: The Army's Way of War* (Cambridge, MA: Harvard University Press, 2007); Roger W. Barnett, *Navy Culture: Why the Navy Thinks Differently* (Annapolis, MD: Naval Institute Press, 2009).

29. Alan R. Millett claims that "some weapons even showed national characteristics" because militaries have different performance priorities. For example, the US Army lacked tanks with sufficient firepower in World War II. "Patterns of Innovation in the Interwar Period," in *Military Innovation in the Interwar Period*, ed. Williamson Murray and Alan R. Millett (Cambridge: Cambridge University Press, 1996), 345.

30. Dominick Graham, *Against Odds: Reflections on the Experiences of the British Army, 1914–45* (Basingstoke, UK: Palgrave Macmillan, 1998), 5.

31. Clausewitz, *On War*, 141.

32. Jacqueline Newmyer, "The Revolution in Military Affairs with Chinese Characteristics," *Journal of Strategic Studies* 33 (August 2010): 483–504.

33. Thomas G. Mahnken makes this point in *Technology and the American Way of War Since 1945* (New York: Columbia University Press, 2008).

34. Lynn, *Battle*, xvii–xix.

35. Edward N. Luttwak, *Strategy: The Logic of War and Peace*, rev. ed. (Cambridge, MA: Harvard University Press, 2001), xii; Gray, *Strategy Bridge*, 65.

36. See Martin van Creveld, *The Culture of War* (New York: Ballantine, 2008); and Richard Ned Lebow, *A Cultural Theory of International Relations* (Cambridge: Cambridge University Press, 2008).

37. Victor Davis Hanson, *The Western Way of War: Infantry Battle in Classical Greece* (London: Hodder and Stoughton, 1989); Victor Davis Hanson, *Why the West Has Won: Carnage and Culture from Salamis to Vietnam* (London: Faber and Faber, 2001); Lynn, *Battle*. Hanson's argument is broadly endorsed in John Keegan, *A History of Warfare* (London: Hutchinson, 1993), but is flatly rejected in John France, *Perilous Glory: The Rise of Western Military Power* (New Haven, CT: Yale University Press, 2011), 5; and Johnson, *Afghan Way of War*, 5–6.

7 THE CHINESE WAY OF WAR

Andrew R. Wilson

ANY DISCUSSION of the Chinese way of war must strike a balance to avoid reducing thousands of years of history to a simple formula or assembling diverse military experiences to argue against any such concept. Moreover, it is necessary to understand the basis of comparisons between the Chinese approach to war and that of other nations. Many considerations of a Chinese, Indian, or Japanese way of war emphasize problematic differences with Western ways of war.[1] Obviously, the more simplistic and reductionist the mirror, the less clarity and nuance in the reflected image. Oversimplification can easily arise when national ways of war are compared.

THE GREAT WALL MYTH

Chinese soldiers and statesmen as well as foreign experts often claim that China has a defensive strategic culture and that all the wars in its imperial past involved suppressing internal rebellion or defending territory. A defense white paper was poetic on this point:

> During the course of several thousand years, loving peace, stressing defense, seeking unification, promoting national unity, and jointly resisting foreign aggression have always been the main ideas of China's defense concept. The defense policy of New China has carried forward . . . excellent Chinese historical and cultural traditions.[2]

According to this myth, Chinese strategic culture, which is founded on the offensive and imperial expansion, is the antithesis of the Western tradition. This conventional wisdom is typically focused on the Great Wall as the embodiment of defensiveness, while more sophisticated accounts of the inherently defensive Chinese strategic culture depend on classical texts. This view draws on the Confucian antipathy toward military classics, especially *The Art of War* (Sunzi bingfa). Yet the underpinnings of Chinese strategic culture—perhaps more accurately the tradition of strategic debate—are more complex and appealing than contemporary caricatures, and the relevance of the Great Wall is almost entirely ahistorical.

China fought many wars to defend territory seized from its neighbors. In addition, wars in the classical age were primarily campaigns of conquest, not only beyond the Sinic zone but also within it. Most iconic Chinese philosophers and strategic thinkers accepted that their civilization would be shaped in the future by peripheral kingdoms that benefited from expansion rather than by geographically limited states along the Yellow River. Only states such as Qi in the north, Qin in the west, and Chu in the south could mobilize the men and matériel to play the role of unifier. Even though harmony might have been an ideal in the core, expansionism on the frontiers was the enabler of imperial consolidation. Instead of avoiding war and expansion, Chinese classics offer abundant rationales to justify the conduct of offensive military operations.

Understandably, justifications for expansion at the expense of non-Chinese peoples were more abundant, explicit, and far less problematic, as seen in *Ancient China and Its Enemies* by Nicola Di Cosmo.[3] Examples of the histories commonly linked to the early Confucians—the *Spring and Autumn Annals* and the *Zuo zhuan* (*The Zuo commentary*)—deal extensively with warfare, and the Han-era chronicle *Shiji* (*The historical records*) notes 483 wars in earlier histories. The Confucian tradition of war often is regarded as a manifestation of and warning against aggression and imperial hubris. However, wars also are sources of many rationales for subjugating non-Chinese peoples. Confucianism thus presents mixed messages to military leaders and rulers of classical China, as it insists that the Son of Heaven is obligated to march against any individual who does not acknowledge his authority.

Within the Sinic zone even Confucians and Daoists finally resigned themselves to wars of conquest and annexation. Although latter-day disciples of Confucius never abandoned the ideal that moral excellence was required for political supremacy, reality forced them to concede that that alone was not sufficient. New forms of political and military organization in the Warring States became so efficient at consolidating resources that even a moralizer like Mencius was forced to acknowledge that a king might be as virtuous as the ancients, but if he did not reign over a state with ten thousand chariots, he could not become a Son of Heaven.[4] The Huang-Lao School of Daoism and Legalists of Qin were more direct in their views: "Thus when the sage attacks and annexes another's state he tears down their walls, burns their bells and drums, disperses their stores, scatters their sons and daughters, divides their territory in enfeoffing [under a feudal estate] the able, this is known as 'Heaven's achievement.'"[5] In other words, the principal tenets of Chinese classical thought included expansion, subjugation, and annexation.

During the imperial era (221 BC–AD 1912) the question of war versus expansion shifted from the contest for internal hegemony to debates over the nature and applications of Chinese power. At this time the argument was focused on the relationship between the state and commerce, with some regarding trade as a legitimate government function to enrich the public treasury. This led to new trade routes to places like Korea, Turkestan, and Vietnam. However, for ideological reasons, some wanted state funding to come exclusively from tax on agriculture. Within a certain school of orthodox Confucian thought, as laid out in the Salt and Iron Debates of the early Han, linking public finance to commerce and expansionism was anathema because it promoted greed when the state was expected to be associated with the better instincts of human nature.[6] Debates on this dilemma continued into modern times, proving that, despite imperial orthodoxy, disputes over expansionism and differing concepts of security have remained intense.

The Great Wall is also deeply problematic as a historical symbol of Chinese defensiveness. The long bulwarks of the Spring and Autumn and Warring States periods, from which the initial Great Wall was strung together by Qinshi Huangdi, were as much about delineating and holding conquered territory as defending traditional frontiers.[7] Long walls served as expedient logistical channels along which men and supplies moved to support territorial conquest and expansion. To impute the defensive purpose of later grand fortifications to these earlier walls both ignores history and overlooks the restless expansionism that characterized many Chinese states. Even the iconic Great Wall of the Ming, a barrier against the Mongols, represented the height of Ming expansion rather than the dividing line between Chinese civilization and steppe barbarism.[8] The Great Wall seen from the north reveals the power that created it. By analogy, Hadrian's Wall does not represent the modest or defensive inclinations of Imperial Rome.

Chinese imperial expansion always has been rooted in security and prosperity. The polities of China expanded in support of ever-larger populations and built buffers around the cultivated core. In that sense, little separated ancient agrarian states. The control and expansion of territory was bound up in the legitimacy of the dynasty. An inability to control or expand territory hinted at a decline in imperial power that was emblematic of inexorable decline. The Communist Party and Chinese people have inherited an obsession for territorial integrity, while many other nations are becoming less preoccupied with sovereignty. With its growing military prowess, fixation on national territory, and insecurity over sovereignty, this could represent a grave danger for China and its neighbors. But fortunately opposition to adventurism as embodied by neo-Confucianism and the Chinese history of imperial overreach may serve as cautions.

Yet it is ironic that the territories over which Beijing makes some of its most strident claims to sovereignty and the right of defense—Tibet, Taiwan, and Xinjiang—were incorporated into the Chinese body-politic through the force of arms. Taiwan was seized from the Netherlands by a Chinese corsair in the 1660s and integrated by the Qing after the Battle of Penghu in 1683. The Qing progressively occupied Tibet over the course of the eighteenth century, at about the same time that the Manchus conquered what is euphemistically known as the Xinjiang Uighur Autonomous Region in China of today.[9]

THE SUN TZU MYTH

One of the most quoted passages from the Sun Tzu (Sunzi) can be translated as "The acme of skill is to win without fighting."[10] This appealing military precept seems to correspond to the injunction by Sun Tzu that military superiority is confirmed by attacking enemy strategies and alliances and inferiority by attacking armies and cities. But appreciating that maxim requires historical context. Sun Tzu first appeared in the Warring States era (403–221 BC) in the context of the rise of states, which were large and powerful enough to compete for mastery over China. As war became more lethal, it became more expensive. This cast the military as the guarantor of wealth and power and also as the biggest drain on resources. Thus, winning without resorting to costly battle contributes to the reason for having the military, which is enhancing the state's wealth and power.

This passage is also directed at anachronistic notions about the purpose of war and the nature of military leadership. Although new technologies and forms of organization were being deployed on the battlefield, the warrior aristocracy of the earlier Bronze Age still monopolized the source of power and aristocratic values prevailed. Two qualities were paramount to nobles: gentility of birth and valorous conduct. Spurning battle ruled out any display of martial virtue and social status and subverted another cause for taking up arms—spilling enemy blood to honor one's ancestors. Thus, the argument by Sun Tzu makes battle not an end but a means to an end.

If winning without fighting was the dominant prescription of the text, one might expect to find more on how to realize this lofty aspiration. A lot in *The Art of War* is about deception, maneuver, and overawing opponents, but far more is on actually winning the battle itself. War is a good deal more violent and purposeful than implied here. Sun Tzu did not eschew battle but, rather, sought to guarantee that the commanders of armies—the greatest affair of the state—be selected for the right reasons. Only rulers who employed a general who followed Sun Tzu principles stood a reasonable chance of success in the zero-sum Warring States contest. Their forces had two tasks: maintaining a credible threat of destruction and actually destroying things. Sun Tzu himself, the putative author of *The Art of War*, was famous for winning.[11] Subjugating an enemy without resort to battle might best be seen therefore as an ideal rather than practical advice.[12]

Although the practicality of winning without fighting does not hold up to much scrutiny, the myth endures that it is a core tenet of the Chinese way of war. The oft-cited examples of winning without fighting are the Chinese propensity to use soft power over hard power: to buy off would-be invaders with gold and women and to practice deft diplomacy or use "barbarians to control barbarians." In cases of actual invasion, the answer was to overcome foreign invaders, like the Manchus, through Sinicization. One overturns foreign conquest by culturally conquering the invader.[13] True, examples of these types of stratagems abound, but so too do examples of the successful use of force by Chinese states.

Arthur Waldron's essay in *The Making of Strategy* and Alastair Iain Johnston's *Cultural Realism* go the furthest to demolish the myth that China is dismissive of the use of force.[14] Nowhere is this more true than in the case of Mao Zedong, whom Samuel Griffith regards as the most notable disciple of Sun Tzu.[15] For Mao, revolution is an act of violence by which one class overthrows another. Thus, revolutionary warfare involves the effort to unseat the dominant class, which has significant material and military advantages that will not be given up without a battle. For Mao, the idea of winning without fighting was inconceivable.[16]

If winning without fighting is regarded as an ideal, one must recognize that *The Art of War* established that the concept of warfighting represented only a means to an end and not an end in itself early in the evolution of Chinese strategic thinking. It also encouraged considering the use of force in holistic and strategic terms. Although similar notions were pondered in the West, the canon of strategic thought did not coalesce in Europe until at least the Renaissance.[17] China, therefore, had a significant lead in strategic analysis and critical thinking.

The concept of *winning without fighting* has emerged in contemporary Chinese literature in reference to legal warfare targeted at promoting specific interpretations of international law and passing domestic laws favoring Chinese interests. In other words, China is attempting to strengthen its legal position, especially regarding territorial claims. Legal warfare strategies have been prevalent in the disputes over the East and South China Seas where sovereignty is contested, jurisdiction overlaps, and international law remains open to debate.

With regard to its exclusive economic zone, China is pursuing strict interpretation of the control of foreign activities there to claim the right to interdict or disrupt previously legitimate activities of other nations. Legal warfare is intended to handicap competition and allow China to take the strategic initiative and moral high ground. Thus, it provides an unassailable legal position that compels the opposition to desist without a legal or military fight.[18] Winning without fighting is not uniquely Chinese. In fact, although the Chinese approach to legal warfare is surely more offensively minded, it is not that radically different from the way the United States uses law and interpretations of law as a strategic instrument.

THE GOOD IRON MYTH

The abject humiliation of China in the nineteenth and twentieth centuries convinced many people that the Chinese lacked a martial ethos. The adage that good men are not used for soldiers, as good iron is not used for nails, undergirded this explanation of military weakness.[19] Keeping the best men, good iron, for *wen* (civil) tasks, like staffing the imperial bureaucracy, meant that China tended to be well governed, but the lack of good men for *wu* (martial) tasks laid it open to aggression from more martial civilizations.

The disparaging tone of the good iron adage also speaks to a perception that the martial arts have historically been disdained by the civilian elites. The early bureaucratization of Chinese states also seemed to imply that from a very early point there was little need for a hereditary Chinese warrior class. As with our other myths, the veracity of the good iron idea does not hold up to much scrutiny.

First, from the Bronze Age to the tenth century at least, the Chinese warrior class competed and collaborated with the civilian bureaucracy. As Ralph Sawyer so skillfully argued, rather than a patent dichotomy between *wen* and *wu*, *wendao* and *wudao* (the civil way and the martial way) are subsets of the larger *wangdao* (kingly way).[20] For example, *The Art of War* was just as much an assault on preachy Confucian moralists as an argument against the pretensions and privileges of aristocrats in the Warring States period. It attempts to strike a balance between two extremes. *Medieval Chinese Warfare, 300–900* by David Graff chronicles the centrality of martial values throughout the Tang, the most indigenous of the imperial dynasties.[21]

One might concede that *wen* won out once Confucianism, and especially the decidedly antimartial neo-Confucianism, became imperial orthodoxy in the Song Dynasty. But that was not the case. The martial continued in symbiosis with the civil.[22] For example, the Song waged two wars against the Tanguts of the Ordos region in the eleventh century. The first Sino-Tangut War of 1038–1044 saw 1.25 million Chinese troops mustered against 826,000 Tanguts. In a second war in the 1080s, some 1 million Song soldiers penetrated deep into Tangut territory before their elaborate logistical train was cut by Tangut cavalry, and the Song forces retreated after suffering 600,000 casualties.[23] As for disdain of soldiers by civilian officials, examples abound, particularly in the late Ming Dynasty when the neo-Confucian bureaucracy rebelled against the military adventurism and pro-military inclinations of the Emperor Wanli, but these were far more battles over policy matters than deeply ingrained cultural tendencies.

Someone enamored of the good iron myth could argue that the *wu* matters most in founding dynasties, when one needs more good men to be soldiers, but then the balance gradually shifts to the *wen*. Dynasties endure because they allow civilians to gain ascendance, or as characterized by one observer, China might have been conquered on horseback, but it has never been ruled from horseback. Even though that is true, there were examples of martial vigor in the Ming and Qing Dynasties. During the Korean campaign of 1592–

1598, the Ming armies were equipped with standard firearms and augmented by artillery, cavalry, and naval forces. The campaign was complicated by logistics.[24] In the nineteenth century during the Taiping Rebellion and Moslem Rebellions, commanders such as Zeng Guofan and Li Hongzhang, who were graduates of the *civil* service examinations, rather than the military examination system, were thus steeped in the orthodoxy of neo-Confucianism and yet proved more than up to the tasks of organizing and leading armies.[25]

With the good iron myth thoroughly tarnished, what glint of the true Chinese way of war might we discern? Even early Chinese states followed a model more similar to the *strategoi* of Athens than did the warrior kings of Sparta. This might have provided rulers and officials with more flexibility in shifting back and forth between civil and military roles. That the Communist Party and People's Liberation Army (PLA) are so closely entwined might be viewed as a continuation of this tradition as much as it is a holdover from Lenin and Mao. The subordination of both the *wendao* and the *wudao* to the *wangdao* further means that the Chinese were better prepared to transition from war to peace and back. Finally, Chinese states traditionally valued bureaucratic aptitude at least as highly as military skill, which explains their impressive military. They have been larger in area and population and far more bureaucratic than Western nations, and they have been able to fight wars on a scale inconceivable in the West until modern times. Therefore, China had greater experience with the strategic, operational, and logistical issues that have been the obsession of Western statesmen and commanders since the eighteenth century.

It could be argued that the supposed dichotomy between the civil and the martial in China has been turned on its head. *Civil-military integration*, much in vogue in the last decade, refers to creating a joint civil-military security system that leverages both civilian and military expertise to the causes of national defense and national modernization. And yet for all the rhetoric and periodic successes, such as Chinese counterpiracy deployments, at the level of national policy and national security strategy there is a deep and growing gap between civilians and the military personnel who staff the bureaucracies of the government, especially the Ministry of Foreign Affairs. The PLA, with its direct link to senior communist leadership, is more influential than the civilian ministries. In terms of foreign policy, particularly when it comes to coordinating national responses that cross jurisdictional lines, military and civilian organizations are rarely on the same page and often oppose each other. This disconnect between the *wen* and the *wu* in China—in which belligerent PLA hawks regularly drown out more conciliatory civilian voices—introduces a significant challenge to the development of constructive relations between China and its neighbors.[26]

THE ZHENG HE MYTH

As recently as 2011, an official of the Ministry of Foreign Affairs explained to a group of foreign academics that China is not expansionist by nature and that one need only look at the remarkable voyages of Admiral Zheng He during the Ming Dynasty as evidence. To quote this Chinese diplomat, Zheng He proves that "the Chinese have no expansionism in their DNA."[27] During the last decade, the Zheng He narrative has been evoked as a popular metaphor for China's peaceful rise and principally for its growing maritime power.[28] It has been pointed out that China in the fifteenth century built the largest ships in the world, the medieval equivalent of aircraft carriers, but did not use them for gunboat diplomacy or in seizing colonies. As China continues to evolve as a maritime power, according to this myth, it will follow the tradition of Zheng He and not the example of Western and Japanese navies. Even though this is an attractive story line, it bears little resemblance to the actual voyages of Zheng He.[29]

Zheng was a eunuch retainer of the Emperor Yongle and rose to prominence as a military commander and head of palace servants. His unique role as senior eunuch and comrade-in-arms made him an obvious choice to lead voyages to the Western Ocean. His seven voyages fall into three groups. One group comprised the first (1405–1407), second (1407–1409), and third (1409–1411) voyages, which opened the Strait of Malacca and revived contact with traditional trading partners in the Indian Ocean. Another group includes the fourth (1412–1415), fifth (1417–1419), and sixth (1421–1422) voyages, which expanded both diplomatic and commercial relations with the Middle East and East Africa. A seventh expedition (1431–1433) retraced earlier trips as far as the Strait of Hormuz and sent out smaller contingents to East Africa. In addition to the distances covered on the voyages, the scale and the likely expense of the enterprise was astounding. A voyage could involve some 250 ships and twenty-seven thousand personnel. Between 40 and 69 of these vessels were treasure ships, the largest perhaps displacing more than twenty thousand tons. Although impressive, the fleet was not designed to engage other navies but rather to overawe enemies or, barring that, disembark large numbers of troops.[30]

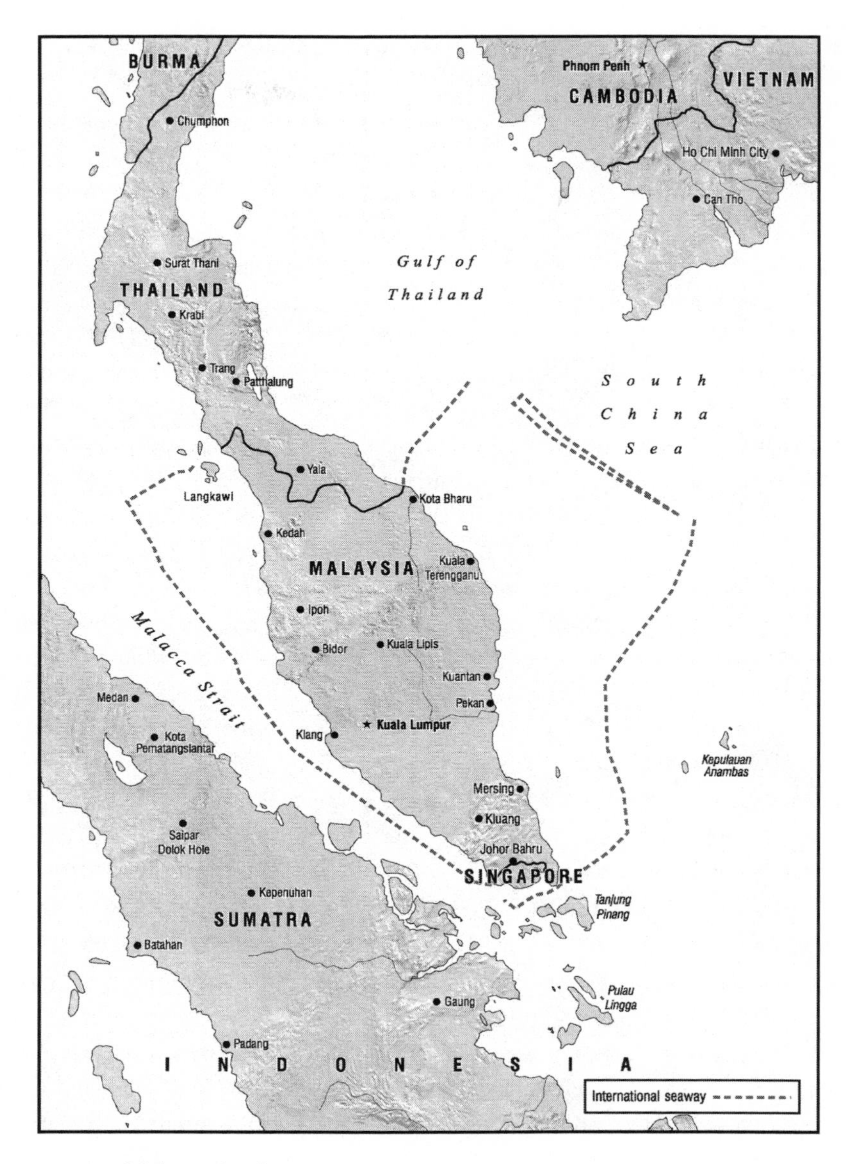

Map 7.1. Malacca Strait.

The boundaries of the Malacca Strait as defined by the International Hydrographic Organization extend on the west from northern Sumatra to the southern coast of Phuket Island in Thailand, on the east from the southern end of the Malay Peninsula to Klein Karimoen Island off the coast of Indonesia, on the north from the southwestern coast of the Malay Peninsula, and on the south from northeastern Sumatra as far as Klein Karimoen. The Malacca Strait is 500 miles long and 155 miles wide in the north and 40 miles wide in the south. (The irregular lines demarcating the boundaries of the Malacca Strait in the Gulf of Thailand and South China Sea indicate [1] a joint development area based on resource claims made by Malaysia and Thailand and [2] overlapping continental shelf claims as defined by Malaysia and Vietnam.)

The expeditions were not seeking new lands or markets as European voyages of discovery did, because they followed well-established routes with diplomacy and information being the main purposes, though the military aspects cannot be denied. Yongle and Zheng He were soldiers and the voyages were organized as military operations. Zheng He was not averse to violence, which was inflicted on the pirate Chen Zuyi at Palembang on the first voyage and the king of Ceylon on the third.[31] In addition, expeditions called at the main port of Champa because it was allied with Yongle in his war against Vietnam. Thus, the strategic aspects of the voyages are apparent. Neither Yongle nor Zheng He distinguished between the civil and the martial spheres of action. Instead, they saw all instruments of Ming power as of one piece.[32]

With regard to colonization, the gravitational pull of the Chinese market attracted trade, as opposed to the European market, which sought control of key trading hubs. Zheng He sought to advertise the glories of the Ming Dynasty and expand the axes along which its gravitational pull could be felt. The objective was not to colonize maritime Southeast Asia or dominate the Indian Ocean but to recruit client states that would recognize the unique and superior status of the Chinese emperor. At the same time, many of the ports visited were already Chinese colonies in an economic sense, and ethnic Chinese ruled at least one of them, Palembang. Finally, although Zheng He was not a colonizer, Yongle was at the time attempting to conquer and annex Annam (northern Vietnam).

In conception and execution, the Zheng He voyages were almost the polar opposite of the Zheng He myth as still floated in Beijing, but even if we dismiss the voyages as an aberration in Chinese maritime history, the myth of no overseas colonies and no gunboat diplomacy still does not hold water. After a period of relative neglect, the Ming returned to the sea with a vengeance in the mid-sixteenth century. In response to piracy and smuggling, Ming civil and military officials, notably Hu Zongxian and Qi Jiguang, developed innovative measures to combat piracy. Their success on land morphed into maritime operations against offshore bases and pirates at sea. By the late 1560s China was deploying naval vessels armed with cannon to suppress the piracy.[33] In the 1570s Ming naval forces pursued Lin Feng to Luzon and collaborated with Spain to expel the Chinese corsair from Lingayen.[34] And during the Imjin War in the 1590s, Ming naval forces bought the Chinese and their Korean ally precious time to build up the ground forces required to drive the Japanese off the peninsula. Moreover, maritime power meant that the

Japanese warlord Toyotomi Hideyoshi could not seize Taiwan, attack Chinese ports, or harass Chinese trade as he had planned. During and after the war, the Ming also were not opposed to threatening retaliation against neighbors who might side with Japan or act against Chinese interests.[35] Finally, in the 1660s, Zheng Chenggong, the Ming proxy known as Koxinga, drove the Dutch out of Taiwan and threatened Manila.[36] In other words, China has been quite willing over the centuries to use gunboats in war and peace.

The arrival of Europeans in East Asia sparked Chinese emigration. By the beginning of the seventeenth century there were Chinese enclaves in Batavia, Japan, Luzon, Malacca, and Taiwan. These entrepôts were hybrid colonies with small numbers of Europeans, twenty thousand to thirty thousand traders and craftsmen from China, and unskilled indigenous populations. Hence, formal colonization was unnecessary. Taiwan was an exception, but its annexation was by no means certain. For the initial three decades of the Qing Dynasty there was little that the Manchus could do to force the Zhengs out of Taiwan. In the 1680s, a former Zheng lieutenant named Shi Lang used bribery, amnesties, and military operations to get the Zheng forces to capitulate. In addition, he successfully lobbied for the formal annexation of Taiwan, a move that the Ming had never attempted but that was viewed as critical for Qing coastal security. In the end, annexation was facilitated by Taiwan having a large ethnic Chinese population thanks to the Dutch.[37]

Although not as aggressive as the Europeans and Japanese of the nineteenth and twentieth centuries, Imperial China still has a modest history of gunboat diplomacy, which can be discerned from the Chinese history of naval warfare, and of overseas colonization, which can be discerned from the history of trade between China and Southeast Asia. But Imperial China lacked the economic and strategic rationales to make these two missions centerpieces of a maritime strategy. Why was this the case? First, near-seas missions, such as coastal defense and counterpiracy, have been Chinese tradition since at least the fourteenth century. Second, despite the Zheng He voyages, far-seas operations were not practical missions for Chinese naval forces. Foreign navies conducting such missions better served Chinese security. Finally, even within the first island chain, colonization was rare except when Chinese interests were threatened or local authorities proved too weak to govern. In more recent years, however, the People's Republic of China has not been averse to using naval force as an instrument of foreign policy. These contemporary versions of gunboat diplomacy are both a welcome development and a cause for concern.

First, the dozen-plus PLA Navy (PLAN) deployments to the Gulf of Aden to support counterpiracy operations are welcome. These symbolic far-seas operations are valuable opportunities for other navies to engage with the Chinese and for the Chinese to realize how far it must go to become a credible blue-water force.[38] Equally welcome have been the goodwill missions by the PLAN hospital ship *Peace Ark* and PLAN support to expatriate evacuation from Libya.[39] This manner of gunboat diplomacy presented China as a responsible stakeholder in the international community. Of concern is the belligerent employment of Chinese forces, in the form of coast guards and armed fisheries vessels, in territorial disputes in the South and East China Seas. The threatened use of force by Beijing to reinforce unrealistic territorial claims is unproductive for maintaining regional stability and relations with its neighbors. In the long term, gunboat diplomacy works against Chinese national interests.[40]

THE MYTH OF *SHI*

Of the five myths about the Chinese way of war, the importance of *shi* is the least problematic in historical terms. *Shi*, variously translated as "strategic advantage," "strategic configuration of power," and "propensity and momentum," to name a few, is in fact a core concept in the classics of the Warring States period from which it seeped into other works on topics such as politics, logic, statecraft, and painting. In *The Art of War*, *shi* makes its appearance in the first chapter: "Assess relative advantage by means of net assessment and use this to achieve *shi* in order that you might use it beyond your borders."[41] In this context, *shi* derives from manipulating the scales (or balance among the factors of *dao* [or *tao*], weather, terrain, command, and doctrine) to maximize strategic advantages against an enemy. Elsewhere in *The Art of War*, *shi* is compared to a drawn crossbow and mass of water released from great height. Obviously, it is a relative and contingent gauge of the conditions found on the battlefield inflicted by both tangible factors (terrain, numbers, and weaponry) and intangible factors (morale, surprise, and deception).

At first glance *shi* appears to be a fairly straightforward military concept that Julius Caesar as well as the emperor of China would understand. Some, however, have pointed to *shi* as a point of fundamental difference between East and West and between the Chinese and Western ways of war. A sweeping work titled *The Propensity of Things: Toward a History of Efficacy in China* by François Jullien elevates *shi* to a central place in Chinese epistemology. Jullien defines *shi* as a nonteleological explanation of outcomes based on relative configurations at a unique time and place and the intrinsic tendency of a situation to evolve along a particular course. In the military realm this equates to a uniquely Chinese way of war that is utterly alien to Western conceptions of cause and effect and theory and practice.[42] Moreover, a brilliant translation of *The Art of War* by Roger Ames adopts a similar interpretation of *shi* to argue that the doctrine of immanence in ancient Chinese epistemology conditioned strategists to wait before determining the relative *shi* of an engagement rather than attempt, in a Western fashion, to control it in advance by means of elaborate planning or the consideration of different strategic theories.[43]

Both Ames and Jullien have produced thought-provoking works, and much can be said in favor of seriously interrogating whether *shi* is a true dividing line between East and West. But embracing this interpretation of the meaning and importance of *shi* inclines students of Chinese military thought and practice to default to the most abstract or mystical explanations of events. When the meaning of *shi* is analyzed, especially in *The Art of War*, things become less abstract and less alien. The concept of *shi* evolved over the course of Chinese history into an abstraction, but when it comes to *The Art of War* and other military classics, *shi* sounds like something that either Caesar or Clausewitz might consider in assessing the surprising ways that discipline, morale, terrain, timing, chance, and genius combine in battle.

Another study that depends on the trope of *shi* as uniquely Chinese is *The Philosophy of Chinese Military Culture* by William Mott and Jae Chang Kim, who are premature in claiming the use of *shi*-based strategies during the twentieth century. They also claim that the employment of the military by Mao in the Civil War, Korea, and subsequent border disputes is inexplicable in Clausewitzian terms. This is ironic given that Mao was deeply influenced by Clausewitz.[44]

The embrace of the explanatory power of *shi* by Mott and Kim is intriguing, but it is also deeply flawed. First, it assumes that Chinese strategic culture and Chinese culture writ large are monolithic and hegemonic, a notion that others have already demolished. Second, it ignores the impact that non-Chinese ways of war, notably German and Soviet, have had on Chinese leaders. Mao might have been a product of a traditional background and steeped in popular culture, but he was also a Marxist and devout Leninist. This had a profound effect on his strategic views. What is more, Mao was influenced by Clausewitz, having studied *On War* in the 1930s together with his lieutenants.

Lenin also borrowed from Clausewitz, which reinforced Mao's direct reading of the Prussian theorist. Although Mott and Kim fail to see Clausewitz in modern Chinese warfare, Mao had no trouble explaining strategy in Clausewitzian terms. Finally, several claims by Mott and Kim, such as the generalization that Communist forces won the Civil War without fighting, in accord with the Sun Tzu myth, have been thoroughly disproved.[45]

If there is something to *shi* in explaining the Chinese way of war, it is the propensity of Chinese strategic theorists and military historians to see god in the details. At numerous points in *The Art of War*, the general is equated to god and his soldiers to mindless automatons. The ideal general thus has godlike abilities, distancing him from mere mortals in the uniqueness of the skill sets involved and the extent of the specialized knowledge required to command. Rather than trying to plumb the divine through omens, a Sun Tzu general becomes divine and imposes order on the infinite complexity of war. Over the course of Chinese military history, this context-specific invocation of godlike powers has evolved into a general notion that in war the general is the puppet master, capable of plumbing and manipulating the complexities of the engagement. David Lai notes that Chinese strategy and grand strategy are explicable in terms of the board game *wei-qi*, known as *Go* in Japan.[46] That approach might have some utility, but it tends to reduce war to a mental contest between opposing generals. War is ultimately a contest of wills, but ignoring Clausewitzian fog, friction, and chance may lead to a dangerous confidence on the part of some Chinese that their strategic culture gives them an edge in controlling an engagement and, especially, escalation.

One venue in which *shi* received much attention in recent years has been cyber-warfare. The appeal of *shi* to Chinese cyber-theorists is quite understandable. Developing cyber-*shi* is a low-cost and clandestine way to prepare the battlefield: secretly outmaneuver an opponent and get the formations in place—computer viruses, malware, and spyware. As a result, clever cyber-combatants are prepared to release a torrent of *shi* to blind and render an opponent unable to communicate with subordinates. Borrowing a line from Sun Tzu, cyber-*shi* allows one to win the battle before it is joined. This is particularly appealing to Chinese cyber-theorists because they cannot go toe to toe with many potential adversaries, especially the United States. But because those potential adversaries depend on information technology for command and control, that creates vulnerabilities that Chinese cyber-warriors can exploit.[47] Even better, the cyber-domain is global and persistent because it is a venue in which an adversary can build up *shi* in advance of a conflict. Thus, when a conflict erupts, a nation like China does not have to mobilize a cyber-army and march deep into enemy territory; it can suddenly unleash all of that pent-up cyber-*shi*. Information technology is evolving so rapidly that even a rising power can develop outsize strategic capabilities within the cyber-domain.

This apparent embrace of cyber-*shi* is potentially bad for China and its neighbors. It is bad because excessive faith in decisive cyber-attack short-changes a more balanced development of the Chinese military. In addition, though cyber capabilities may be emerging as a great strength for China, the more it becomes the centerpiece of PLA strategy and military operations, the more of a vulnerability those capabilities will become. If Beijing places too much faith in cyber-attack, it also will be bad news for its neighbors. If China is convinced that it has a war-winning stratagem, a way to quickly seize control of the terrain and climate of cyberspace, it may wage what some analysts call a highly controlled war—a war in which one side dominates in all of Sun Tzu's categories: weather, terrain, command, doctrine, and information.[48] But such warfare is rare, and that level of control is usually short-lived. In the last analysis, thinking that war is something that can be highly controlled is dangerous.

LOOKING AHEAD

After reviewing the five cardinal myths with more historical accuracy than they usually receive, the question remains as to which aspects of the Chinese military tradition still inform its strategy. One thing seems clear: there is no single Chinese way of war. One must take history and culture seriously when considering strategic inclinations, and as scholarship of the past two decades on Chinese military history and culture clearly has shown, Chinese military traditions are rich and diverse. But a Chinese strategist today who believes that the future trajectory of Chinese geostrategy is determined by an intrinsic Chinese way of war is no more correct in evaluating tradition than an ancient Confucian who predicted that an empire may be united only by a sage king. Today, Chinese political and military leaders are free to choose different models of military action in formulating policy, and though culture is one resource they may draw on, it does not confine their actions to a predictable range. Even though interest in the classics and the great captains and past campaigns is growing in China, the move back to tradition is tentative and superficial.

Most of what appears in the Chinese media relies on tropes demolished above, and scholarship on the mainland is not particularly sophisticated.

Many specialists in Chinese security studies appear to be obsessed with strategic culture and unearthing the Chinese way of war. This is an attempt to find both greatness and solace in history. Although a Chinese way of war existed in the past, tradition has been under intense attack for most of the last century, and the worldview of Chinese leaders has been so conditioned by foreign ideologies and military doctrines that it is hard to find much continuity with that past. Aspects of contemporary Chinese strategy resonate with the past but may be coincidental. A deeper understanding of how statecraft has been practiced in peace and war is needed, but in the near term one should not be surprised by the continuation of self-Orientalizing reductionism that has characterized the debate.

Understanding Chinese military experience calls for examination of the counterpoints to the myths that matter, including territory, size, strategy, culture, and the near seas. China has waged countless wars of territorial expansion and defense. Because controlling continental and maritime arenas is intimately linked to political legitimacy, China has tended to overemphasize the value of territorial objectives. The large Chinese population, relative economic prosperity, and sophisticated bureaucracy led to gigantism in military affairs. From vast continent-spanning fortifications and treasure fleets to the herculean efforts of the Korean War, even recent Chinese governments have translated immense military potential into immense military capability. While quantity may have a quality of its own, the gigantism of the past and present China has produced important works of strategy, among them *The Art of War* and writings of Mao Zedong. Perhaps it is because of the scale of warfare that thinking about the linkage between military operations and their political consequences has enjoyed a rich tradition in China.

• • •

Historically, China has been a continental power, but it has been putting to sea since at least the tenth century. Proximity to the ocean and the first island chain is vital to Chinese security and prosperity. For most of its long history China has dealt with a region marked by both uncertainty and overlapping sovereignties. That worked to the advantage of Chinese states, because as long as a hostile power did not control the near seas, they enjoyed the fruits of maritime trade without the need for a large navy. Today Chinese views are conditioned by a century of humiliation. If given the requisite military capability and provocation, the Chinese will go to war at sea.

As opposed to a single coherent strategic culture, myths represent flawed assumptions that many Chinese and foreign observers enthusiastically embrace. Myths not only seek to define a Chinese way of war but posit alterities that define the strategic culture of a potential enemy. The *yin-yang* of the philosophy of Daosim and the dialectics of Marxism-Leninism may explain contemporary Chinese strategists who regard the inclinations of adversaries as the polar opposite of their strategic culture myths. China is not alone in embracing this version of strategic polarity, but given the insecurities that plague its growing military potential, it is nonetheless troubling to find questions of policy, strategy, and war dealt with in simplistic ways.

NOTES

1. Victor Davis Hanson, *Carnage and Culture: Landmark Battles in the Rise to Western Power* (New York: Doubleday, 2001); John A. Lynn, *Battle: A History of Combat and Culture* (Boulder, CO: Westview, 2003).

2. People's Republic of China, "China's National Defense" (white paper, Information Office of the State Council, Beijing, July 1998).

3. Nicola Di Cosmo, *Ancient China and Its Enemies: The Rise of Nomadic Power in East Asian History* (New York: Cambridge University Press, 2002).

4. *Mencius*, trans. D. C. Lau (New York: Penguin, 1970), 75–76.

5. Mawangdui Hanmu boshu zhengli xiaozu [馬王堆漢墓帛書整理小組; Mawangdui Han Tombs Silk Text Compilation Team], *Mawangdui Hanmu boshu* [馬王堆漢墓帛書; *Mawangdui Han Tombs Silk Text*] (Beijing: Wenwu, 1980), 45.

6. Sadao Nishijima, "The Economic and Social History of Former Han," in *Cambridge History of China*, vol. 1, *The Ch'in and Han Empires, 221 BC–AD 220*, ed. Denis Twitchett and John King Fairbank (Cambridge: Cambridge University Press, 1986), 545–607.

7. Di Cosmo, *Ancient China*, 127–58.

8. See in particular Arthur Waldron, *The Great Wall of China: From History to Myth* (Cambridge: Cambridge University Press, 1990).

9. Peter C. Perdue, *China Marches West: The Qing Conquest of Central Eurasia* (Cambridge, MA: Harvard University Press, 2005).

10. All translations from the Sun Tzu are the author's. For a one of the better published translations, see Sun Tzu, *The Art of War*, trans. Samuel B. Griffith (New York: Oxford University Press, 1971). The relevant passage is chapter 3, verse 3, on page 77 of the Griffith version.

11. Andrew Meyer and Andrew R. Wilson, "*Sunzi Bingfa* as History and Theory," in *Strategic Logic and Political Rationality: Essays in Honor of Michael I. Handel*, ed. Bradford A. Lee and Karl Walling (London: Frank Cass, 2003), 99–118.

12. Some argue that the ideal-real, or theory-practice, dichotomy reflects Western conceit and was not shared by the ancient Chinese. See Francois Jullien, *A Treatise on Efficacy: Between Western and Chinese Thinking* (Honolulu: University of Hawaii Press, 2004).

13. On the Sinicization thesis, see Ping-ti Ho, "In Defense of Sinicization: A Rebuttal of Evelyn Rawski's 'Reenvisioning the Qing,'" *Journal of Asian Studies* 57 (February 1998): 123–55.

14. Arthur Waldron, "Chinese Strategy from the Fourteenth to the Seventeenth Centuries," in *The Making of Strategy: Rulers, States, and War*, ed. Williamson Murray, Alvin Bernstein, and MacGregor Knox (Cambridge: Cambridge University Press, 1996), 85–114; Alastair Iain Johnston, *Cultural Realism: Strategic Culture and Grand Strategy in Chinese History* (Princeton, NJ: Princeton University Press, 1998).

15. Sun Tzu, *Art of War*, 45–56.

16. Mao Tse-tung, "On Protracted War," in *Selected Military Writings of Mao Tse-tung* (Peking: Foreign Languages Press, 1967), 2:143–44.

17. Beatrice Heuser, *The Evolution of Strategy: Thinking War from Antiquity to the Present* (Cambridge: Cambridge University Press, 2010).

18. Dean Cheng, "Winning a War Without Fighting," *Washington Times*, July 17, 2013, http://www.washingtontimes.com/news/2013/jul/17/cheng-winning-a-war-without-fighting.

19. See Morton H. Fried, "Military Status in Chinese Society," *American Journal of Sociology* 57 (January 1952): 347–57.

20. Ralph D. Sawyer, "Martial Prognostication," in *Military Culture in Imperial China*, ed. Nicola Di Cosmo (Cambridge, MA: Harvard University Press, 2009), 45–64.

21. David Graff, *Medieval Chinese Warfare, 300–900* (New York: Routledge, 2001), 160–251.

22. Kathleen Ryor, "Wen and Wu in Elite Cultural Practices During the Late Ming," in Di Cosmo, *Military Culture*, 219–41.

23. See Paul C. Forage, "The Sino-Tangut War of 1081–1085," *Journal of Asian History* 25 (1991): 1–28.

24. Kenneth M. Swope, *A Dragon's Head and a Serpent's Tail: Ming China and the First Great East Asian War, 1592–1598* (Norman: University of Oklahoma Press, 2009).

25. Maochun Yu, "The Taiping Rebellion: A Military Assessment of Revolution and Counterrevolution," in *A Military History of China*, ed. David A. Graff and Robin Higham (Boulder, CO: Westview, 2002), 135–51.

26. The PLA has become sharply critical of conciliatory positions advocated by the Ministry of Foreign Affairs. See Michael D. Swaine, "China's Assertive Behavior,

Part Four: The Role of the Military in Foreign Crises," *China Leadership Monitor* 37 (Spring 2012): 1–14; and Alice Miller, "The Political Implications of PLA Professionalism," in *Civil-Military Relations in Today's China*, ed. David M. Finkelstein and Kristen Gunness (Armonk, NY: M. E. Sharpe, 2006), 131–45.

27. Anonymous official of the Chinese Ministry of Foreign Affairs, interview by the author, June 28, 2011. As recently as April 2014, the Chinese Ministry of Foreign Affairs described the deep roots of China's resistance to expansionism: "Peace and cooperation, pronounced 'He' in Chinese, is a principle and code of conduct deeply rooted in the traditions of the Chinese culture. 'He' implies peaceful, harmonious and auspicious; it can also refer to cooperation, integration and combination. This cultural DNA affects the behavior pattern of the Chinese people as well as the way China handles its relations with its neighbors. Over 600 years ago, Zheng He, a Chinese navigator, made seven expeditions to the western seas with a fleet which was then the world's strongest. Unlike Western colonists who engaged in expansion and plunder, Zheng He's fleet brought Chinese silk, porcelain and tea to the countries it visited and left behind touching stories of friendship between the Chinese and locals. The concept of 'He' featured prominently in this episode of history." Embassy of the People's Republic of China in Malaysia, "Promote Cooperation, Manage Differences in the Spirit of 'He,'" April 24, 2014, http://my.china-embassy.org/eng/zgxw/t1149709.htm.

28. The series in China's official navy newspaper to commemorate the six hundredth anniversary of Zheng He's voyages offers a recent example. See, for instance, 徐起 [Xu Qi], 敦睦友邻—郑和下西洋对中国和平崛起得启示 [A friendly neighbor promoting friendly relations: The inspiration of Zheng He's voyages to the West in China's peaceful rise], 人民海军 [People's Navy], July 12, 2005, p. 3.

29. The best account in English of the voyages is Edward L. Dreyer, *Zheng He: China and the Oceans in the Early Ming Dynasty, 1405–1433* (New York: Longman, 2007), 238–85.

30. Ibid., 23–26.

31. Ibid., 53–60, 79–81.

32. Andrew R. Wilson, "The Maritime Transformations of Ming China, 1360–1683," in *China Goes to Sea: Maritime Transformation in Comparative Historical Perspective*, ed. Andrew S. Erickson, Lyle J. Goldstein, and Carnes Lord (Annapolis, MD: Naval Institute Press, 2009).

33. Kenneth M. Swope, "Cutting Dwarf Pirates Down to Size: Amphibious Warfare in Sixteenth-Century East Asia," in *New Interpretations in Naval History: Selected Papers from the Fifteenth Naval History Symposium*, ed. Maochun Yu (Annapolis, MD: Naval Institute Press, 2009), 81–107. For more on Qi Jiguang, see Ray Huang, *1587, a Year of No Significance* (New Haven, CT: Yale University Press, 1981), 156–88.

34. "Lin Feng," in *Dictionary of Ming Biography*, ed. L. Carrington Goodrich and Chaoying Fang (Cambridge: Cambridge University Press, 1976), 1:917. See also 陳荊和 [Chen Jinghe], 十六世紀之菲律賓華僑 [Shiliu shiji zhi feilupin huaqiao; The Philippine-Chinese in the sixteenth century] (Hong Kong: Southeast Asian Studies Section, New Asia Research Institute, 1963), 31–41.

35. Swope, *Dragon's Head*, 114–21, 250–51.

36. Lynn A. Struve, "The Southern Ming, 1644–1662," in *The Cambridge History of China: The Ming Dynasty, 1368–1644, Part I*, ed. Frederick W. Mote and Denis Twitchett (Cambridge: Cambridge University Press, 1988), 663–76.

37. Tonio Andrade, "Pirates, Pelts, and Promises: The Sino-Dutch Colony of Seventeenth-Century Taiwan and the Aboriginal Village of Favorolang," *Journal of Asian Studies* 64, no. 2 (2005): 295–320; Tonio Andrade, *How Taiwan Became Chinese: Dutch, Spanish, and Han Colonization in the Seventeenth Century* (New York: Columbia University Press, 2006).

38. Andrew S. Erickson and Austin M. Strange, *No Substitute for Experience: Chinese Antipiracy Operations in the Gulf of Aden* (Newport, RI: China Maritime Study Institute, Naval War College, 2013).

39. Gabe Collins and Andrew Erickson, "China Dispatches Warship to Protect Libya Evacuation Mission: Marks the PRC's First Use of Frontline Military Assets to Protect an Evacuation Mission," *China SignPost* [洞察中国] 25 (February 24, 2011): 1–3.

40. Michael Richardson, "China's Gunboat Diplomacy, *Japan Times*, July 30, 2012.

41. Author's translation; the relevant passage appears on page 66 of the Griffith translation.

42. Francois Jullien, *The Propensity of Things: Toward a History of Efficacy in China* (New York: Zone, 1999).

43. Sun Tzu, *The Art of War*, trans. Roger T. Ames (New York: Ballantine Books, 1993), 39–63.

44. William H. Mott IV and Jae Chang Kim, *The Philosophy of Chinese Military Culture* (New York: Palgrave MacMillan, 2006), 1–44.

45. On the military aspect of the Civil War, see Odd Arne Westad, *Decisive Encounters: The Chinese Civil War, 1946–1950* (Stanford, CA: Stanford University Press, 2003).

46. David Lai, "Learning from the Stones: A Go Approach to Mastering China's Strategic Concept, *Shi*" (Carlisle Barracks, PA: Strategic Studies Institute, US Army War College, 2004).

47. Bryan Krekel, Patton Adams, and George Bakos, "Occupying the Information High Ground: Chinese Capabilities for Computer Network Operations and Cyber Espionage," US-China Economic and Security Review Commission, March 7, 2012; Timothy L. Thomas, "China's Cyber Tool: Striving to Attain Electronic *Shi*?" in *Law,*

Policy, and Technology: Cyberterrorism, Information Warfare, and Internet Immobilization, ed. Pauline C. Reich and Eduardo Gelbstein (Hershey, PA: IGI Global, 2012), 409–27.

48. Steve Winterfeld and Jason Andress, *The Basics of Cyber Warfare: Understanding the Fundamentals of Cyber Warfare in Theory and Practice* (New York: Syngress, 2012), 40.

8 THE JAPANESE WAY OF WAR

S. C. M. Paine

THE SOCIAL SCIENCES routinely assume the rational actor to investigate human behavior. However, culture can trump rationality. As this chapter demonstrates, culture had an enormous impact on the way the Japanese waged World War II. Their way of war in the Pacific Theater continues to mystify Americans. Why did many choose suicide over surrender? Why did they continue to fight at such enormous personal cost long after hope of winning had disappeared? For Americans such choices make no sense. But different values predispose different choices. The Japanese conduct of war is a cautionary tale for those who make predictions based on an assumed monopoly on rationality, which they define in terms of the universality of their own values. An examination of Japanese literature can help explain their choices.

TRADITIONAL THEORISTS

Japanese military literature is founded on a warrior ethic. The code of the samurai (*bushido*) is the counterpart to Carl von Clausewitz or Sun Tzu (Sunzi), comprising a body of literature that focuses not on strategy but on military deportment, a subject generally absent in Western sources. This emphasis reflects social values that put a premium on proper decorum and the meticulous respect for complex rules of etiquette. The most famous books on this topic were produced by retired samurai in the seventeenth and eighteenth centuries, during the Tokugawa period (1603–1867), which ironically became well known for its domestic tranquility.

A more recent book can serve as a cultural bridge to understand the Tokugawa literature. Nitobe Inazō's *Bushido: The Soul of Japan* (1900) concisely defines *bushido* and examines its philosophical origins. At the time of World War II, it was one of the most widely read books on military thinking and was published in eight languages.

According to Nitobe, Japanese ideas on right and wrong came not from religion but rather from *bushido*, which he defined as the Precepts of Knighthood, a code of honor for the warrior class.[1] The three pillars of *bushido* were Buddhism, Shinto, and Confucianism.[2] From Buddhism came the Japanese sense of fatalism that contrasts with free will, a major theme in Western philosophy. Buddhism also provided the basis of the Japanese attitudes toward death. Their belief system includes four noble truths: existence is suffering; craving and attachment cause this suffering; there is an end to suffering, nirvana; and there is a path toward that end in the form of right conduct. Notably, the emphasis is placed on deportment, on means and not ends, and on the way people lead their lives and not what they achieve. From Shinto came extreme loyalty to the emperor and deep patriotism. According to Shintoism, emperors descend from the sun goddess in an unbroken line. From Confucianism came ethical doctrines governing a hierarchical society with interlocking social obligations.

The West upheld rights, while the East respected duties. The West promoted equality, and the East stressed hierarchy. Traditionally, equality did not figure as an Eastern social category; relations were always between superiors and inferiors. Even identical twins were not equals; rather, birth order made one the elder and superior. The combination of etiquette and ritual controlled relations among unequals. This emphasis on deportment contrasted with both the Greco-Roman canon and Judeo-Christian beliefs. These differences have implications. People of different cultures prioritize values and measure costs and benefits in their own way. The American emphasis on life, liberty, and the pursuit of happiness differs from loyalty, duty, and honor, which constitute the foundations of *bushido*.

Yamamoto Tsunetomo (1658–1719) wrote the most famous book of Tokugawa literature, *Hagakure* (literally "hidden leaves"; 1716), which contains the thoughts of a close retainer of a feudal lord in southwestern Japan. However, Yamamoto never participated in battle; the values that he espoused applied to conditions of a century earlier.

Preoccupation with Death

Western readers of samurai literature are struck by its obsession with death. While Clausewitz stressed violence in war, he did not concern himself with what constitutes an honorable death. By contrast, Yamamoto reports, "the way of the samurai is, morning after morning, the practice of death, considering whether it will be here or be there, putting one's mind firmly on death."[3] He found that "merit lies more in dying for one's master than in striking down the enemy" and that, "pressed with the choice of life or death, it is not necessary to gain one's aim," but to live after failure is cowardice. Thus, if the samurai "is able to live as though his body were already dead, he gains freedom in the Way. His whole life will be without blame."[4]

Note that Yamamoto emphasizes devoted service and loyalty over strategic effectiveness, not Clausewitz's focus on the achievement of the policy objective. In the West, dying for one's cause might, perhaps, be a strategy, but it would not normally constitute an objective in the way that it became one for many Japanese during World War II.[5] Consider the banzai charges, when Japanese remnants chose being cut down by machine guns over surrendering.

Honor and Suicide

Japanese ideas on death have been intimately associated with honor. According to Yamamoto, "Victory and defeat are matters of the temporary force of circumstances. The way of avoiding shame is different. It is simply in death. . . . A real man does not think of victory or defeat. He plunges recklessly towards an irrational death."[6] In the event of a shameful failure Yamamoto endorses suicide: "The least he could do [is] to cut open his stomach, rather than live on in shame." Otherwise, one's family would be disgraced.[7] Yamamoto is referring to social death. Returning from war defeated brings eternal shame to oneself and one's family, descendants, ancestors, and fellow samurai. His solution is hara-kiri (*seppuku*), which entails disemboweling oneself with a short sword, followed by decapitation by one's second, usually a friend. Although this practice appalled Western sensibilities, Nitobe claimed it was a means of death "associated with instances of noblest deeds and of most touching pathos, so that nothing repugnant, much less ludicrous, mars our conception of it. . . . [T]he vilest form of death assumes a sublimity and becomes a symbol of new life."[8] Hara-kiri was failing with honor.

Emphasis on Loyalty

Honor is connected to loyalty. According to Yamamoto, "Being a retainer is nothing other than being a supporter of one's lord, entrusting matters of good and evil to him, and renouncing self-interest. If there are but two or three men of this type, the fief will be secure."[9] Service to one's lord or company comes before family, unlike in China, where family comes first.

These ideas concerning death, honor, suicide, and loyalty possess operational and strategic implications. The Tokugawa literature assumes an objective of operational success focusing not on what to do beforehand to prepare the ground for victory but on what happens afterward in the event of failure. The answer is not the Western approach of damage limitation; rather, it is a particular kind of damage limitation in terms of family honor and not in terms of the physical costs of defeat. Suicide preserves honor in the event of failure. In war there are two possibilities: to win or die trying. In World War II, such beliefs meant the Japanese soldiers were not going to surrender even when confronted by overwhelming superior forces.[10]

Westerners often assume enemies are rational actors who make cost-benefit calculations to determine when to cut their losses. Although this may apply to the West, it does not project well across cultural boundaries to civilizations where costs and benefits may be calculated differently. Consider Japanese veterans discovered after World War II holding out alone in remote jungles.[11] As Field Marshall Sir William Slim, who led the Fourteenth Army in Burma, observed, the Japanese soldier "fought and marched . . . till he died. If five hundred Japanese were ordered to hold a position, we had to kill four hundred and ninety-five before it was ours—and then the last five killed themselves."[12] Another British veteran who fought in the same theater called Japanese troops "the bravest people I have ever met. . . . They believed in something, and they were willing to die for it, for any smallest detail that would help achieve it."[13]

Emphasis on Willpower

Honor, suicide, and loyalty converged in willpower. According to Yamamoto, "No matter what it is, there is nothing that cannot be done." He goes on, "The way of the samurai is in desperateness. Ten men or more cannot kill one such man. Common sense will not accomplish great things. Simply become insane and desperate." If one is "unattached to life and death . . . one can accomplish any feat."[14] Think of the young kamikaze pilots, the supreme expression of honor, suicide, loyalty, and willpower. The premium on willpower differs from the call by Clausewitz for reassessment when the costs far exceed the value of the objective.

Denigration of Strategy

Emphasis on willpower degrades the importance of strategy. As Yamamoto argues, "Learning such things as military tactics is useless," and he recommends reflexively striking at the enemy. The way of the samurai calls for immediately rushing headlong at the enemy.[15] He elaborates, "There is a saying that goes, 'No matter what the circumstances might be, one should be of the mind to win. One should be holding the first spear to strike.'"[16] Doubts paralyze those versed in military tactics, so act quickly; do not deliberate. While Yamamoto writes about tactics and not strategy, his work has strategic implications and takes an unanalytical approach to warfare. He finds that studying the past is unnecessary to derive lessons: instead, maintain correct deportment and steel the will for victory. Absent are grand strategy, alternatives, and interaction. For example, before the Japanese entered Indochina in 1940, they did not consider the international consequences of their advance. They steeled the will for victory and advanced, which precipitated the Allied oil embargo and the Japanese reaction at Pearl Harbor.[17]

This dismissive attitude toward strategy and emphasis on the will over material limitations combined to produce the tremendous optimism in Japanese planning throughout World War II. Moreover, it helps explain the recklessness with which Japan took territories without considering the ultimate costs, let alone the ability to exploit indigenous resources.

Taira Shigesuke (1639–1730), known as Daidōji Yūzan, was a contemporary of Yamamoto, a Confucian scholar, and a military strategist. He described samurai values in the *Code of the Samurai* (1730): "On the warrior's path, only three things are considered essential: loyalty, duty and valor." This entailed loyalty in defeat because the "warrior is still fiercely loyal and does not leave his employer's side even when a hundred allies are reduced to ten, even when ten are reduced to one, so steadfastly loyal in battle as to disregard his own life."[18] This meant being loyal to the group through the fulfillment of duty to the group. An in-group–out-group dichotomy pervades Japanese society. While individuals constitute the basic unit of society in the West, and Westerners often speak of the rights of individuals, the basic unit in Japan is the group, and group interests take precedence over individuals and personal interest.[19]

There are many categories of group membership. Japanese constitute the overarching in-group vis-à-vis the outside world. Groups are divided by region, workplace, and units within organizations, schools, and families. Each person belongs to numerous overlapping in-groups and owes them varying degrees of loyalty and potentially competing obligations. The grammar of the Japanese language reflects hierarchical gradations, with group members addressing one another in more personal or familiar terms. When Japanese speak, grammar and word choice make hierarchical relationships clear: who is superior and who is inferior as well as whether they belong to the same group.[20] Thus, loyalty to the group is far more deeply embedded in the Japanese population for cultural, social, and linguistic reasons.

Miyamoto Musashi (1584–1645) was a masterless samurai who lived as a swordsman and teacher of martial arts. According to his *The Book of the Five Rings* (1643), "To fight even five or ten people single-handedly in duels with your life on the line and find a sure way to beat them is what my military science is all about. So what is the difference between the logic of one person beating ten people and a thousand people beating ten thousand people?"[21] In other words, long odds did not faze the Japanese. In the Russo-Japanese War (1904–1905) and Pacific War (1941–1945), the Japanese military leadership estimated the prospect for success to be fifty–fifty or less in both instances, and still they recommended going to war.[22]

Miyamoto offered advice on breaking a military stalemate, which was the problem during the Second Sino-Japanese War (1931–1945). "If there is a total deadlock . . . it is essential to stop right away and seize victory by taking advantage of a tactic unsuspected by the enemy."[23] In other words, use a new tactic in pursuit of the original objective. The Japanese often did that by opening new theaters with preemptive attacks, a tactic favored by Miyamoto.[24] The First Sino-Japanese War, Russo-Japanese War, Second Sino-Japanese War, and World War II in the Asia-Pacific region all began with preemptive attacks. Japan responded to the stalemate with campaigns in Manchuria in 1931, Inner Mongolia in 1933, the Chinese seaboard and Yangzi River Valley in 1937, and multiple theaters in the Pacific in 1941.[25]

Miyamoto provides advice on breaking the will of an enemy, which was the problem that faced the Japanese in China. He comments on a defeated adversary who refuses to capitulate or the modern problem of defeating a Fabian strategy or an insurgency: "You can knock the heart out of people with weapons or with your body or with your mind. . . . When your enemies have completely lost heart, you do not have to pay attention to them any more."[26] This might have been the intent of the Japanese atrocities in the Second Sino-Japanese War.

The Tokugawa literature reveals a preference for offensive strategy, audacity, and preemption. When outcomes fail to meet expectations, look for daring tactics in pursuit of the original goal. Such values imply a rigidity: if unsuccessful, try harder. With inflexible standards for success, reassessment is regarded as a failure in the original strategy, not as a response to fog, friction, and chance. Writers of the Tokugawa period stressed doing things right the first time without backup plans, let alone branches and sequels. Because they thought in terms of operational success and not policy goals, they did not consider strategic alternatives.

The Japanese remained fixated on all-or-nothing scenarios during World War II, which, once decided on, had to be rigidly followed. In the event of failure, the exit strategy was suicide. Indeed, the stress on honor made an exit strategy almost inconceivable. Those who failed had suffered social death, which suicide recognized. The moment they started to fail in battle, they were on *death ground*, as Sun Tzu described the situation of those trapped and for whom fighting offered the only hope of survival.

In World War II, the Japanese did study strategy in their war colleges. They read works by foreign military experts, but older preferences lingered. They focused on strategy at the tactical and operational levels, discounting grand strategy.[27] By doing so, they committed the fatal sin of omission. Unlike in the Meiji period, they lacked a grand strategy encompassing military, political, economic, diplomatic, and also cultural components. They saw a range of threats and maintained an expanding list of ambitions, which together forced them to pursue strategic opportunities without a clear definition of victory. Instead of deciding how much territory to occupy before declaring victory, their territorial objectives became a function of operational success and strategic failure. The inability to pacify China led Japan to look further afield. Moreover, they had no strategy for war termination. In the Meiji period civil and military leaders jointly conducted war termination, but in the Great Depression, the military muzzled their civilian counterparts.

SAMURAI VALUES IN PRACTICE

In World War II, four areas were particularly revealing of *bushido* in action: choices concerning logistics, sea lines of communication, and intra- and interservice rivalries.

Neglect of Logistics

The samurai literature stresses the importance of willpower and ability of will to trump material limitations, an emphasis that led to defeat in World War II.

The Japanese industrial base was comparatively small. Their steel and coal production were about one-thirteenth of the US output, and munitions production never exceeded 10 percent.[28] There were four tons of equipment for each American fighting in the Pacific; two pounds for each Japanese soldier. The Imperial Army was equipped with obsolete weapons, especially artillery, machine guns, and tanks. Although every infantryman in the US Army had eighteen men supporting him in combat and every soldier in other Western armies had eight men supporting him, Japan reportedly had a one-to-one tooth-to-tail ratio.[29]

Even before Pearl Harbor, the Japanese home islands suffered from food shortages. By the early fall of 1942, fuel oil had become scarce, and the winter of 1942–1943 saw an acute shortage. This meant the fleet could no longer maneuver at will. Energy shortages decreased the output of steel and aluminum.[30] Thus, as US factories started to gear up production, Japanese industry began to fall apart. In early 1945, the Japanese estimated they would run out of aviation and motor fuel by the end of the year.[31] Yet they remained opposed to surrender. The samurai code endured: fight on despite long odds, remain loyal, persevere, and never give up.

On August 15, 1945, the day Japan capitulated, Admiral Ugaki Matome, who had been the chief of staff of the Combined Fleet and at the end of the war organized kamikaze attacks, noted in the final entry in his diary, "There are various causes for today's tragedy, and I feel that my own responsibility was not light. But, more fundamentally, it was due to the great difference in national resources between the two countries."[32] This realization came too late. While American production statistics had been published for decades, many in the Japanese military had regarded them as propaganda because they seemed ludicrously high. Those officers who had lived in the United States or Great Britain and knew better were considered defeatist and not promoted.[33] Ugaki was a practitioner of *bushido* as well as a believer in the ability of willpower to overcome material limitations. The war proved his optimism to be ill founded.

Neglect of Sea Lines of Communication

The last premier of Imperial Japan, Prince Higashikuni Naruhiko, told the Diet in late 1945 that "the basic cause of defeat was the loss of transport shipping."[34] Japan had occupied countries in the Pacific to secure resources, but they were useless unless Japan could ship them home. But the Imperial Navy had given little consideration to the problem of protecting the necessary sea lines. Instead, it focused on the symmetrical fight of main fleet engagements.

A shortage of shipping capacity developed in the winter of 1942–1943. By war's end, Japan had one-ninth of its December 1942 merchant marine capacity.[35] Not only did the Japanese fail to protect merchant shipping; they did not target US activity such as lend-lease cargoes bound for the Soviet Union.[36] Not until the battle of Saipan (June–July 1944) did Admiral Ugaki start recommending a more Fabian strategy. By then Japan lacked the ships for successful fleet-on-fleet engagements and the resources to attack enemy shipping. In early 1942, he dismissively entered in his diary, "It is too bad for the officers and men of the submarine service that they have not yet sunk any important man-of-war, only merchantmen."[37]

On the first anniversary of Pearl Harbor, while trying to explain the failure at Guadalcanal, Ugaki wrote, "The aim of supply and transportation to the front has not been even half-fulfilled each time. It led those on the verge of death to be extremely skeptical about the Navy. . . . They considered it selfish, only seeking its own ends at the sacrifice of the Army, using the latter as its decoy."[38] Many army officers believed the navy discounted supplying expeditionary forces, and they considered convoy and transport duties as essential for success.[39] It turned out willpower could not compensate for limited industrial capacity or the challenges of defending huge geographic areas. Loyalty, duty, and honor predisposed the Japanese to discount such limitations.

Intraservice Rivalries

On the surface, the fracture lines of loyalty—namely, service, branch, year of entry, and unit—may seem similar to the fracture lines in other militaries. Yet these divisions were much more extreme in Japan, where the degrees of membership and hierarchy within groups were far more finely calibrated. These fracture lines were inherently more difficult to overcome because such factors as in-group–out-group divisions reinforced administrative divisions at the root of service rivalries. By 1944, the upper echelon of the Imperial Army concluded belatedly that interservice rivalries constituted the greatest impediment to prosecuting the war.[40]

Crosscutting loyalties in the army made unified command difficult and cooperation with the navy highly problematic. For example, the decision in 1931 to invade Manchuria was made by the Kwantung—also known as the Kantōgun or Manchurian—Army rather than by Tokyo. The leaders of the

Kwantung Army believed that they had the clearest understanding of national interests. Consequently, they unilaterally invaded and occupied Manchuria, an area larger than France and Germany combined.[41] Throughout the Second Sino-Japanese War, army and navy factions assassinated civil and military leaders. When the emperor decided to capitulate, the army attempted one final coup, which luckily for ending the war failed.

The Imperial Navy also suffered from dysfunctional intraservice divisions. Japan could not replace experienced pilots lost in the war. Unlike the US military, the Japanese did not alternate combat and training assignments. The pilots signed up together, trained together, fought together, and died together. They belonged to groups differentiated by year of induction into the service. In-group affiliations did not preclude but certainly inhibited drafting seasoned pilots into other groups to train younger pilots.[42] Because the survival rate of experienced pilots was low and there was an unwillingness to admit failure, the Japanese did not easily learn from tactical and operational mistakes and improve their performance. By contrast, US failures such as the defense of Pearl Harbor drew headlines and were vigorously debated.

Interservice Rivalry

In-group–out-group divisions influenced all levels of the Imperial Army and Navy commands. The fracture lines got much worse between the services, which had become adversaries in the prewar budget battles. The differences escalated in tandem with the overextension of Japanese forces. Interservice rivalry manifested itself in administrative dysfunction, mutually exclusive war planning, and reciprocal deception. Throughout war, the army and navy proved unable to collaborate on any level. In 1944, under threat of imminent defeat, they attempted to coordinate strategy but managed only to schedule liaison conferences. In 1945, the army sought to unify service high commands but combined only information departments.[43] Even with the threat of invasion the services failed to merge their capabilities to defend the home islands.[44] Had they coordinated aircraft production, standardized parts could have kept more planes in combat.[45] This lack of military harmonization was endemic throughout the war.

From 1906 on, the Japanese Army and Navy were allowed to develop completely separate war plans that did not even focus on the same enemy, yet each service assumed the other would provide key support.[46] Although the Imperial Army drew up plans to fight the Soviet Union, the so-called northern advance plan, the navy concentrated on fighting the United States and Britain for the Empire in the Pacific, or the southern advance plan. The army belatedly came around to the naval plan after Russia made the northern advance a nonstarter by decimating Japanese forces on the Mongolian border during the Battle of Nomonhan in 1939.[47]

Nevertheless, during the first five months of 1942, Japanese forces took more ground over a greater area than any other military in history. And the navy did not lose a single major ship.[48] The Japanese mind-over-matter strategy seemed to be effective. The army immediately started to redeploy troops back to the China theater, where it planned to commit 250,000 soldiers out of 450,000 by the end of 1942 to defend against the Russians. However, the navy was planning to advance to New Guinea and Australia.[49] Thus, just as the United States was gearing up for the counteroffensive, Japan was removing its land forces from the region.

The navy, however, neglected to inform the army that it was not actually ready for a big southward advance. Success presupposed a well-fortified outer defensive perimeter from which to fend off an anticipated US counterattack, but the navy had yet to reinforce enough islands to form a strong outer perimeter. It was frantically building airfields, including one at Guadalcanal. The army did not learn of this situation until the day in August 1942 when US troops landed on the island. In response to calls from the navy for help, the Seventeenth Army was ordered to capture Port Moresby in New Guinea and to defend Guadalcanal. In keeping with *bushido*, they did not alter their strategy; they simply tried harder, despite New Guinea and Guadalcanal being more than six hundred miles apart. On Guadalcanal, the army exaggerated the number of frontline troops so the navy would not short them on rations, but the troops starved anyway.[50] In the end, the navy wanted to cut its losses but the army wanted to fight on.[51]

Guadalcanal had enormous strategic implications. Japan had planned the Gogō Offensive up the Yangzi River to eliminate the Chinese Nationalists. When the Japanese defeated them in their capital in Nanjing in 1937, the Nationalist government retreated farther up the Yangzi River and set up its capital at Chongqing, which the Japanese planned to attack in 1942 until Guadalcanal made it impossible.[52] The next stop up the Yangzi for retreat would have been Burma. Without Guadalcanal, the Japanese might well have knocked the Nationalist government out of the war, freeing hundreds of thousands of Japanese troops to fight the United States.[53]

The navy deceived the army again at Midway. Admiral Yamamoto Isoroku had goaded the army into accepting his plan for the battle. He threatened to resign unless the army went along, but it did not respond.[54] The Doolittle Raid in April 1942 successfully bombed several Japanese cities, including Tokyo. The attacks horrified the Imperial Army, which had failed to protect the skies over Japan. The raid was insignificant in operational terms, causing little damage. However, it had enormous strategic ramifications because it induced a blunder. Two days later, the army went along with the Midway plan.[55] Here samurai values can be seen in action: avenge one's honor; do not think, retaliate. At Midway, the US Navy sank four enemy carriers (one-third of the force), a loss that was kept from the Imperial Army and government for several months.[56] The navy informed the emperor, who apparently did not share the information.[57] In this case, in-group loyalty took precedence over national interests.

War Termination

Until the final days of war, the Japanese sought an operational rather than a strategic solution. They focused on fighting harder and not on diplomacy with friends or foes.

The Axis alliance was strategically ineffectual. Germany and Japan had much in common ideologically, and each considered the Soviet Union its primary adversary. Accordingly, both powers worked to hem in the Red Army, but each overextended itself by fighting secondary enemies. Much in keeping with samurai literature, the Japanese repeatedly failed to cut their losses in China and never attempted seriously to negotiate with the United States. They simply fought on ever more brutally. In the end, Japan did get a military solution: unconditional surrender after a war of attrition involving the firebombing of Japanese cities.

Three events that occurred over a four-day period combined to end the deadlock in Tokyo. On August 6, the United States dropped an atomic bomb on Hiroshima. On August 8, the Soviet Union declared war, pouring 1.5 million troops into Manchuria to create the nightmare scenario long feared by war planners. Finally, on August 9, a second atomic bomb fell, on Nagasaki. And even then Japanese capitulation occurred only because of the insistence of Emperor Hirohito, who now feared US bombs and a Soviet invasion of the home islands more than a military coup. On August 10, to break the cabinet deadlock, Hirohito took the unprecedented step of accepting the call by the Potsdam Declaration for an unconditional surrender. On August 15, he delivered a radio broadcast to his people, also unprecedented. On August 16, he sent three imperial princes to the Manchurian, Chinese, and Southern theaters to relay his decision. His samurai then obeyed their lord. From then on the Japanese cooperated with the US occupation forces.

• • •

America achieved its immediate goals in Japan in a very Japanese way. By the end of the war, Washington had come to understand Tokyo, in contrast to its prewar strategy of deterrence by embargo, which backfired. The United States used Emperor Hirohito to secure cooperation by legitimating reform of civil and military institutions. The emperor, who feared that the United States would execute him and end the dynasty, collaborated with General Douglas MacArthur, approving the constitution drafted in one week by the Americans.

In World War II, there were cultural explanations for the Japanese neglecting grand strategy, refusing to cut losses and end the conflict, failing to coordinate operations, and behaving with such ferocity. Culture, however, is not static. Although the suicide rate remains higher in Japan than in other developed countries except for South Korea and Hungary, with three times as many men killing themselves than women,[58] the last notorious case of hara-kiri occurred in 1970. At that time the famous novelist Mishima Yukio killed himself after an impassioned speech in favor of restoring the authority of the emperor to a jeering mob of soldiers in Tokyo.

The tragic events of war changed Japan and its people. Postwar pacifism eclipsed *bushido*. The war had destroyed 66 percent of Japan's national wealth and left sixty-six cities in ruins and 9.2 million people homeless. Only Kanazawa, Kyoto, and Nara survived intact.[59] Japan suffered 28,000 military deaths in 1941, 66,000 in 1942, 100,000 in 1943, 146,000 in 1944, and strikingly, 1.13 million in 1945. Between 3 and 3.5 percent of its population died in the conflict, which took 20 million lives on all sides in the Asia-Pacific region.[60]

In opinion polls conducted in 1993, 2009, and 2010, the postwar constitution, written by MacArthur's staff, remained popular, with roughly two-thirds of Japanese respondents in favor of its retention.[61] Article 9 renounced "war as a sovereign right and the threat or use of force as a means of settling international disputes."[62] The attitude change, reflected in strong public support for Article 9, occurred before 1952, when 45.8 percent already opposed revising the constitution to allow for reconstituting the military and only 27.9 percent supported such change.[63] In a poll in 1953, only 15 percent opposed

Article 9.[64] Just as culture influences the conduct of war, the conduct and outcome of war affect culture.

NOTES

1. Nitobe Inazō, *Bushido: The Soul of Japan* (Boston: Tuttle, 1969), 4, 82.

2. Ibid., 11–15.

3. Yamamoto Tsunetomo, *Hagakure*, trans. William Scott Wilson (Tokyo: Kodansha International, 1979), 73.

4. Ibid., 17–18, 55.

5. Haruko Taya Cook and Theodore F. Cook, *Japan at War: An Oral History* (New York: New Press, 1992), 309.

6. Yamamoto, *Hagakure*, 30.

7. Ibid., 72–73.

8. Ibid., 112, 116.

9. Ibid., 20.

10. John Toland, *The Rising Sun: The Decline and Fall of the Japanese Empire, 1936–1945* (New York: Modern Library, 2003), 723, 734.

11. Hiroo Onoda, *No Surrender* (Tokyo: Kodansha International, 1974).

12. Quoted in Edward J. Drea, *In the Service of the Emperor* (Lincoln: University of Nebraska Press, 1998), 63.

13. Quoted in Richard B. Frank, *Downfall: The End of the Imperial Japanese Empire* (New York: Random House, 1999), 28, 29.

14. Yamamoto, *Hagakure*, 45, 48, 51, 73, 158.

15. Ibid., 159, 60.

16. Ibid., 155.

17. Sadao Asada, *From Mahan to Pearl Harbor: The Imperial Japanese Navy and the United States* (Annapolis, MD: Naval Institute Press, 2006), 259.

18. Taira Shigesuke, *Code of the Samurai* (Boston: Tuttle, 1999), 9, 22.

19. Timothy Hoye, *Japanese Politics* (Upper Saddle River, NJ: Prentice Hall, 1998), 3, 19; Agnes M. Niyekawa, *Minimum Essential Politeness* (Tokyo: Kodansha International, 1991), 25–30.

20. Niyekawa, *Politeness*; John W. Dower, *War Without Mercy* (New York: Pantheon Books, 1986), 234.

21. Miyamoto Musashi, *A Book of Five Rings* (Woodstock, NY: Overlook Press, 1974), 33.

22. Shumpei Okamoto, *The Japanese Oligarchy and the Russo-Japanese War* (New York: Columbia University Press, 1970), 71, 100–101; Alvin D. Coox, "The Pacific War," in *The Cambridge History of Japan*, ed. John W. Hall, Marius B. Jansen, Madoka Kanai, and Denis Twitchett (Cambridge: Cambridge University Press, 1988), 6:329.

23. Miyamoto, *Five Rings*, 40.

24. Ibid., 35.

25. Bruce A. Elleman and S. C. M. Paine, *Modern China* (Boston: Prentice Hall, 2010), 302–5, 318–30; S. C. M. Paine, "Pearl Harbor and Beyond: Japan's Peripheral Strategy to Defeat China," in *Naval Expeditionary Warfare*, ed. Bruce A. Elleman and S. C. M. Paine (London: Routledge, 2011), 70–83.

26. Miyamoto, *Five Rings*, 46.

27. Drea, *In the Service*, 26–27; David C. Evans and Mark R. Peattie, *Kaigun* (Annapolis, MD: Naval Institute Press, 1997), 493–94; David C. Evans, ed., *The Japanese Navy in World War II: In the Words of Former Japanese Naval Officers* (Annapolis, MD: Naval Institute Press, 1986), 502, 512, 515.

28. Coox, "The Pacific War," 377; B. G. Sapozhnikov, *Китайский фронт во второй мировой войне* [The Chinese Front in World War II] (Moscow: Nauka, 1971), 62.

29. Meirion Harries and Susie Harries, *Soldiers of the Sun* (New York: Random House, 1991), 348–51, 369.

30. Ugaki Matome, *Fading Victory: The Diary of Admiral Matome Ugaki, 1941–1945*, ed. Donald M. Goldstein and Katherine V. Dillon (Pittsburgh, PA: University of Pittsburgh Press, 1991), 21, 175, 222, 240–41, 264.

31. Saburo Hayashi and Alvin D. Coox, *Kōgun* (Quantico, VA: Marine Corps Association, 1959), 154.

32. Ugaki, *Fading Victory*, 664.

33. Hayashi and Coox, *Kōgun*, 23.

34. Coox, "The Pacific War," 378.

35. Ibid.

36. Hans-Joachim Krug, Yōichi Hirama, Berthold J. Sander-Nagashima, and Exel Niestlé, *Reluctant Allies: German-Japanese Naval Relations in World War II* (Annapolis, MD: Naval Institute Press, 2001), 76; Asada, *From Mahan to Pearl Harbor*, 266.

37. Ugaki, *Fading Victory*, 66, 418.

38. Ibid., 296.

39. Hayashi and Coox, *Kōgun*, 43, 82; Evans and Peattie, *Kaigun*, 455.

40. Hayashi and Coox, *Kōgun*, 112; Cook and Cook, *Japan at War*, 112.

41. Sadako N. Ogata, *Defiance in Manchuria: The Making of Japanese Foreign Policy, 1931–1932* (Berkeley: University of California Press, 1964), 54, 58–59, 69, 83, 94–97.

42. 安部彦太 [Abe Hikota], "大東亜戦争の計数的分析" [A statistical analysis of the Great East Asian War], in 近代日本戦争史 [Modern Japanese Military History], vol. 4, 大東亜戦争 [The Great East Asian War], ed. 奥村房夫 [Okumura Fusao] and 近藤新治 [Kondō Shinji] (Tokyo: Dōdai Keizai Konwakai, 1995), 4:837; Mark R. Peattie, *Sun-*

burst: The Rise of Japanese Naval Air Power, 1909–1941 (Annapolis, MD: Naval Institute Press, 2001), 107, 131–34, 166–67, 174, 180–84, 191–92.

43. Hayashi and Coox, *Kōgun*, 84–85, 152–53.

44. Harries and Harries, *Soldiers of the Sun*, 450; Coox, "The Pacific War," 369.

45. Hayashi and Coox, *Kōgun*, 80; Peattie, *Sunburst*, 100–101.

46. 黒野耐 [Kurono Taeru], 日本を滅ぼした国防方針 [Japan's disastrous national defense policy] (Tokyo: Bungei Shunjū, 2002), 22–25; 北岡伸一 [Kitaoka Shin'ichi], 日本陸軍と大陸政策1906–1918年 [The Japanese Army and its strategy for the Asian mainland, 1906–1918] (Tokyo: Tokyo University Press, 1985), 9; 角田順 [Tsunoda Jun], 満州問題と国防方針一明治後期における国防環境の変動 [The Manchurian question and national defense policy: Changes in the late-Meiji national defense environment] (Tokyo: Hara Shobō, 1967), 678–700; 大山梓 [Ōyama Azusa], ed., 山県有朋意見書 [Written opinions of Yamagata Aritomo] (Tokyo: Hara Shobō, 1966), 294–300.

47. Alvin D. Coox, *Nomonhan: Japan Against Russia* (Stanford, CA: Stanford University Press, 1990), 914–16; Maochun Yu, *The Dragon's War: Allied Operations and the Fate of China, 1937–1947* (Annapolis, MD: Naval Institute Press, 2006), 19.

48. Robert B. Edgerton, *Warriors of the Rising Sun* (New York: W. W. Norton, 1997), 277.

49. Drea, *In the Service*, 34–36; Hayashi and Coox, *Kōgun*, 45, 47, 50, 65.

50. Coox, "The Pacific War," 357; Evans and Peattie, *Kaigun*, 455; Hayashi and Coox, *Kōgun*, 58, 65.

51. Ugaki, *Fading Victory*, 285–86.

52. 江口圭一 [Eguchi Kei-ichi], 十五年戦争小史 [A short history of the fifteen year war] (Tokyo: Aoki Shoten), 209.

53. 吉田裕 [Yoshida Yutaka] and 纐纈厚 [Kōketsu Atsushi], "日本軍の作戦•戦闘•補給" [Military operations, combat, and the supply of the Japanese military], in 十五年戦争史 [A history of the fifteen year war], ed. 藤原 彰 [Fujiwara Akira] and 今井清一 [Imai Seiichi] (Tokyo: Aoki Shoten, 1989), 3:106.

54. Ugaki, *Fading Victory*, 41; Hiroyuki Agawa, *The Reluctant Admiral Yamamoto and the Imperial Japanese Navy*, trans. John Bester (Tokyo: Kodansha International, 1979), 304; Hayashi and Coox, *Kōgun*, 51.

55. Edgerton, *Warriors*, 275; Drea, *In the Service*, 37.

56. Edgerton, *Warriors*, 281.

57. Herbert P. Bix, *Hirohito and the Making of Modern Japan* (New York: HarperCollins, 2000), 449.

58. US Census Bureau, *Statistical Abstract of the United States: 2011* (Washington, DC: Government Printing Office, 2012), 844.

59. Takafusa Nakamura, *A History of Shōwa Japan, 1926–1989* (Tokyo: University of Tokyo Press, 1998), 25, 219, 245; Michael Clodfelter, *Warfare and Armed Conflicts:*

A Statistical Reference to Casualty and Other Figures, 1618–1991 (Jefferson, NC: McFarland, 1992), 2:951–52; Frank, *Downfall*, 77; S. Woodburn Kirby, M. R. Roberts, G. T. Wards, and N. L. Desoer, *The War Against Japan* (London: Her Majesty's Stationery Office, 1969), 5:102–5, 162, 484–86.

60. 江口圭一 [Eguchi Kei-ichi], 十五年戦争小史 [A short history of the fifteen year war], 172, 226; Nakamura, *Shōwa Japan*, 253–54.

61. Mansfield Asian Opinion Poll Database, "Asahi Shimbun April 2010 Regular Public Opinion Poll" (April 17 and 18, 2010), 10–12; Mansfield Asian Opinion Poll Database, "Asahi Shimbun April 2009 Public Opinion Poll on the Constitution" (April 18 and 19, 2009), 9–11; NHK Broadcasting Culture Research Institute poll, March 1, 1992.

62. Constitution of Japan, chap. 2, "Renunciation of War," art. 9 (November 3, 1946).

63. Allan B. Cole and Naomichi Nakanishi, eds., *Japanese Opinion Polls with Socio-Political Significance, 1947–1957* (Medford, MA: Fletcher School of Law and Diplomacy, Tufts University; Williamstown, MA: Roper Public Opinion Poll Research Center, Williams College, 1959), 3:603.

64. Ibid., 3:607.

9 THE INDIAN WAY OF WAR

Timothy D. Hoyt

TO MANY AMERICANS, India is an alien and, in certain respects, incomprehensible culture. The United States had relatively little interaction with India before the 1940s, unlike with China and Japan. The popular image of India was influenced by motion pictures such as *Gunga Din* and personalities like Mohandas Gandhi. Thus, it is unsurprising that leading American area specialists have sought unsuccessfully to describe Indian culture in terms of permanent attributes, which are manifested in war and politics.[1]

The argument supporting a distinct Indian military tradition is similar to that advanced in the case of the Japanese, American, or Chinese way of war.[2] India possesses an impressive military tradition that is shrouded in mythology and lacking in a recorded history, unlike Western nations. India is a unique civilization that dates back thousands of years, according to myth and traditions. This South Asian nation extends from Baluchistan to the Indo-Burmese frontier and from the Himalayas to the Indian Ocean. Various cultures, religions, and societies have combined to form the Indian-Hindu civilization, which features in the clash of civilizations posited by Samuel Huntington.[3]

THE HYPOTHESIS

The case for an Indian way of war is fundamentally reductionist when compared to Western and Chinese ways of war.[4] India has ancient texts on war, including the mythic *Mahabharata*, which is similar to *The Iliad* or the Old Testament in its fusion of religious and practical reflection, and it also has the historical and didactic *Arthashastra*.[5] In addition, traditional empires located primarily in northern India were periodically reestablished after periods of decline. In the nonimperial eras, however, the Indian political system devolved into competing kingdoms that were similar to the Chinese Warring States or Western Europe after Westphalia. Finally, military traditions in India at the higher levels of warfare were apparently consistent over protracted periods, and therefore, they reflect some of the wisdom and caution of India's ancient texts.

Indian military classics date to pre-Christian times. *The Mahabharata* is a work of myth, but it is also a description of heroic warfare and political intrigue in which both gods and men participate. Similar to *The Iliad* in the Western tradition, it dramatically recounts tales of war, leadership, heroism, and trickery.[6] The Indian armies described in *The Mahabharata* resemble the infantry, cavalry, archers, chariots, and elephants that opposed Alexander the Great during the fourth century BC, although they predate the Greek invasion.[7]

Kautilya wrote *The Arthashastra* as a manual on ruling a kingdom. Its origins are debated, and like *The Art of War*, it might have been written by multiple authors. It is thought to date from between the sixth and third centuries BC, about the time the Mauryan Empire might have emerged in northern India, and is often compared with *The Prince*, by Machiavelli.[8] It implies what has been identified as "an obsession with spies, secrets, and treachery."[9] This preoccupation could be blamed on a predatory international environment similar to that faced by Sun Tzu. *The Arthashastra* can be regarded as a survival manual for weak rulers under internal and external threat who might resort to covert means to increase their power or remain independent. Although not endorsing war, it also presents extensive information on contemporary military affairs. In addition, it emphasizes the uncertainty of war, since the protagonists can employ covert means to disrupt or defeat one another.

Thus, the classics suggest trends in military thinking that have appeared in Indian behavior over several millennia. Such ideas include a generally defensive outlook, a cautious approach to war, and an awareness of the uncertainty of battle. Indian writers are sensitive to internal threats and betrayal and to transitory alliances and coalitions. Winning without fighting is emphasized, and subversion, deception, deterrence, and coercion are favored.

Imperial Heritage

A second indicator of the strong cultural influence on military practice is found in the unified empires and hegemonic kingdoms, particularly those of northern India. During the fourth through the second centuries BC, the Mauryan Empire could mobilize more than six hundred thousand infantry, thirty thousand cavalry, and nine thousand elephants, according to Greek sources.[10] It reached geographic and political heights under Asoka circa 250 BC. Therefore, when Alexander the Great was victorious at the Battle of Hydaspes in 326 BC, he fought a coalition of smaller kingdoms, but he never engaged the expanding Mauryan Empire east of the Indus. Asoka converted to Buddhism after a bloody military contest, and his successors did not enjoy his prowess or success.

Other empires arose in northern India, covering nearly the same area as the Mauryan, although they never fully incorporated the southern portion of the subcontinent. The Guptan Empire controlled much of this territory from the third to fifth centuries AD.[11] The rise of Islam created threats from Afghanistan that established the Sultanate of Delhi in the early thirteenth century.[12] Eventually it was replaced by the Moghuls, who invaded from the northwest and subsequently dominated northern India from 1526 until the late eighteenth century.[13]

These empires unified the rich and populous areas of northern India through conquest and extended their reach southward within the limits of geography and logistics. Each succeeded in resisting external invaders, at least for a while. The Mauryans, as previously noted, profited from a buffer of kingdoms against Alexander.[14] The Guptans drove off the Huns in 457 AD.[15] The Sultanate of Delhi resisted the Mongols for 150 years but succumbed to a brief but ruinous raid by Timur the Lame in the late fourteenth century.[16] The Moghuls ruled for more than two hundred years before their defeat by the Europeans, who came from the sea.[17]

The British Empire expanded slowly, taking almost a century after the Battle of Plassey to control the entire subcontinent.[18] Violent albeit unsuccessful resistance marked this period as the local rulers adapted to European technology and methods. The Marathas, sultans of Hyderabad, and the Sikhs of the Punjab clashed with troops of Great Britain and the East India Company.[19] The British rule highlights the continuity of empires in India, a rule that lasted for almost a century before decolonization established the current regional state system.

Recurring Trends

India does not possess a tradition like the mandate of heaven, but the sequential establishment, expansion, collapse, and replacement of the Mauryan,

Guptan, sultanate, Moghul, and British followed a loose pattern. Each of those empires resisted invasion but eventually succumbed to ineffective governance and declining legitimacy, which led to the loss of allies and supporters and ultimately to military defeats. Even Britain experienced this fate with the formation of the Indian National Army after its humiliating loss to Japan in 1941.[20] In each case, an aggressive though smaller force defeated a larger but relatively passive defender weakened by internal discord that provided the attacker with real or potential allies.

Each new imperial power needed sufficient force to consolidate and govern its conquests. The size of the standing army, even accounting for inevitable exaggeration, was truly formidable and relied on integrating local forces and rulers into the imperial force.[21] As indigenous soldiers were incorporated into a new imperial army, offensive momentum and efficiency declined, and conflicts within the military and among political and religious factions increased. The result was that the army during each period took on a peculiarly Indian character.

Accordingly, India's more recent military history demonstrates some tendencies found in pre-Christian-era Indian armies, suggesting striking continuity. Indian armies were cautious and defensive, relied on mass rather than skill and organization, and minimized risk. Their scope for greater dynamic action was constrained by divisions within the state, society, and military itself. As a result, even the most successful conquerors reached their culminating point of advance and plateaued in terms of both territorial control and political legitimacy. All these experiences accord with classical writings, suggesting cultural traditions manifested in Indian efforts to wage war.

COUNTERARGUMENT

Using selective cultural evidence to avoid contrary arguments could be viewed as reductionist and oversimplified. In fact, the body of Indian literature on military affairs is rather slender.[22] Unlike ancient Greece, Rome, or China, little is known about Indian armies. Reliable accounts only emerged after the triumph of the Moghuls.[23] As a result, specifics on how Indians fought and the quality of their military leadership and training remain scant. Perhaps more is known about the Dark Ages or Celtic world than about India before 1500.

Pre-1500 India clung to traditional forms of warfare long after they became problematic. The chariot, for example, was retained well after other armies abandoned it.[24] Elephants were one constant and were used for millennia, though they were proved ineffective at the Battle of Hydaspes in 326 BC.[25] While the infantry fielded archers, and the cavalry was both large and important, there is little evidence of mounted archers until outside forces led to India's defeat.[26] Although Indians had access to firearms and artillery, they failed to properly utilize them and were surprised by European militaries.[27] Thus, the reliance on mass simply masked the lack of professionalism and inability to derive effective methods of warfare.

Empires

Though it is possible to stress the continuity of Indian imperial rule and practice, each empire fell, and except for the Mauryans, they fell to militarily superior opponents. While this may suggest continuity, it is not evidence for a unique "way of war."

The argument that Afghan, Turkic, and English conquerors were somehow assimilated not only by Indian society but also by the military culture appears to be a substantial overstatement. In fact, religious and racial differences complicated social assimilation, and there is no reason to suggest that military assimilation had any greater success. Each invader employed different, asymmetrical capabilities that proved very successful against traditional Indian methods. Efforts by indigenous forces to adapt or assimilate new methods were unsuccessful until *after* conquest.[28]

The Afghans deployed mounted archers, which were unknown in India. This combination of mobility and firepower overwhelmed more ponderous local armies. The sultanate instituted discriminatory policies rewarding conversion to Islam, undermining the Hindu social order, which hardly indicated successful assimilation by Indian culture. The Moghuls combined horse archers with field fortifications, firearms, and according to reports, artillery at the First Battle of Panipat, achieving roughly the same impact as Nobunaga's army at Nagashino a half century later.[29] The Europeans, especially the English, achieved enormous success when outnumbered through the use of steady, well-trained infantry with artillery support. In each case, the Indians failed to adapt to innovative practices used by an invading force.

More importantly, later Moghul and British empires disempowered Indian military elites, replacing them with imported leadership.[30] The Moghuls depended on officers of Turkish origin, with their entirely different and demonstrably more effective military canon. The English did not allow Indian officers to reach command positions until late in their rule, although they did cooperate with local allied rulers.[31] So the Indian Army of the Raj was actually led, trained, and operated according to British rather than Indian traditions and standards. In addition, the British imposed a racist and somewhat bizarre concept of *martial races* on the Indian forces, depriving themselves of large and talented portions of the Indian population.[32]

Finally, the emphasis on northern empires may simply be convenient as well as misleading. No imperial power before the British, who came from the sea, successfully conquered southern India. These kingdoms and miniempires are not well documented, but they made a real impact on surrounding regions as far away as Indonesia. They resisted northern domination, suggesting the possibility of an alternative Indian military tradition perhaps shaped by geography. The ethnic composition of the southern kingdoms may be an alternative source of military and political traditions that are overlooked by proponents of an Indian way of war.

Recurring Trends

A look at the historical record suggests that at minimum large gaps exist in our understanding of the impact of culture on Indian military traditions. There is a great deal we do *not* know, and proponents of an Indian way of war tend to use culture to bridge these gaps. This assumption is based on the partially successful assimilation of each successive wave of invaders into Indian society—a strong indication of the power of Indian culture across several millennia.

A significant counterargument, however, rests on the most recent imperial experience. It is hard to argue for continuity in the Indian way of war, based on cultural tradition, during the Raj. The soldiers were undoubtedly Indian and, especially during the sepoy period, imposed elements of their culture on the British Army.[33] The officer corps, however, was another matter. It offered a more critical indicator of an Indian way of war because little is known about traditional tactics and practice. And for more than a century, no Indian reached the rank of colonel or commanded a large body of troops, which suggests major discontinuities in military leadership. As a result, it can be argued that Indian military practice is more British than Indian.

SYNTHESIS

It is possible to argue that Indian culture, in particular political culture and tradition, does affect the use and performance of the military. Political culture may initiate predictable *strategic* preferences for the modern Indian state.

But proving that it is an absolute, or that it is based on ancient Indian writings, may be far more difficult to establish.

The modern Indian state is very much a product of the struggle for independence. Its elites were born of that struggle, which was led by barristers such as Jawaharlal Nehru and Mohandas Gandhi. They looked at history through the lens of a conquered people. British imperialism was seen as the result of technologically superior forces and extreme internal divisions, which enabled small armies to defeat large empires and then subjugate the subcontinent.[34] Alliances were considered dangerous, and internal unity was seen as critical to national security.

Indian leaders were also confronted by a military that was the last remaining instrument of the British Empire. Thus, it is hardly surprising that it was regarded as a potential threat to the state. The new regime gave the fledgling naval and air forces equality in stature, downgraded the rank structure of the officer corps, and deprived the army of both resources and respect.[35] Though the degree of distrust has diminished, civil-military relations in India are skewed much more heavily in favor of the civilian sector than in practically any Western nation, and the Indian armed forces thus exercise less influence in the conduct of foreign and defense policy.[36]

Finally, colonialism made Indian elites distrustful of international politics, sympathetic for apparent underdogs, and concerned over the arms race and military aggression.[37] Consequently, India tends to highly respect national sovereignty and rarely undertakes military action unless attacked or involved in United Nations peacekeeping operations. Sometimes these factors result in erratic behavior in modern India. In some respects, it appears to reflect the advice of Kautilya. India is risk averse in foreign affairs and somewhat defensive politically and militarily. It shows great restraint given its forces. It remains cautious about alliances and international ventures and focuses heavily on internal unity and increasingly on economic growth as key sources of power. At the same time, India has had little success in covert operations and is facing internal threats. In this regard Pakistan is probably more Kautilyan than modern India.[38]

In thinking about India it may be worth reflecting on the great work by Russell Weigley, *The American Way of War*, which actually identifies *two* American ways of war, conventional and unconventional.[39] A single Indian way of war may not exist, but one can detect archetypal approaches to conflict in international terms that reflect recurring patterns.

Archetype One: Risk Averse

Indians are reluctant to wage war and have not developed sophisticated means for considering military operations. There are few think tanks and elected officials spend little time on defense affairs. Military planning has been left to professional soldiers operating under the scrutiny of suspicious bureaucrats, and defense budgets have received small allocations of the gross domestic product, which beggars modernization and preparedness.

In addition, India has pursued a largely reactive foreign policy. Most of its wars have been started by adversaries: the invasion of Kashmir by Pakistan in 1947, the military intervention in the Himalayas by China in 1962, the invasion of Jammu and Kashmir by Pakistan in 1965, and infiltration of militants into Kashmir by Pakistan from 1989 to the present, including the border war over Kargil in 1999.[40] Thus, one archetype of Indian warfare is the surprise attack, followed by methodical battle, and slow, constrained, and shallow military operations when penetrating enemy territory. Efforts have been constrained by apparent conservatism, the lack of combined arms and joint force coordination, difficult terrain, and inadequate logistics. Although Pakistan has exhibited risk taking to the point of endangering regional stability, its operational methods mirror those of India. One could argue that both armies demonstrate a fundamental symmetry that is rooted in British training and practice that lasted into the 1950s.

Archetype Two: Risk Taker

Indian national security policy has not always been risk averse. In 1947, for example, Indian and Pakistani accounts differ on when India began airlifting troops to Srinagar to defend Kashmir.[41] Indian intransigence and occasional bellicosity toward China from 1959 to 1962 contributed to Beijing's decision to invade. India also engaged in a prolonged and ultimately unsuccessful peacekeeping mission in Sri Lanka in 1987–1990 and prevented a coup in the Maldives in 1988. India's invasion of Goa in 1961 technically risked war between India and NATO.[42]

Although those were comparatively minor episodes, on at least three other occasions India has taken quite serious risks in pursuit of advancing its regional and global positions. The most recent event involved the decision to conduct nuclear tests in 1998. While India had considered nuclear tests in 1995–1998, they were dissuaded by American diplomacy or domestic politics.[43] But a coalition government led by the Bharatiya Janata Party made testing a priority by carrying out deceptive measures, including reportedly misleading US diplomats, performing a highly publicized missile test to distract US intelligence, and then startling the international community by risking economic sanctions to formally demonstrate its nuclear capability.[44]

India also exhibited considerable risk-taking behavior in 1986–1987. On the western border, New Delhi conducted military exercises of increasing sophistication known as Brasstacks that ended by deploying most of its mechanized formations close to the Pakistani border. This led Pakistan to mobilize its armored forces and caused a standoff that included nuclear threats and India reportedly cancelling Operation Trident, which would have attempted to liberate Pakistan-held Kashmir.[45] India also embarked on a confrontation along the Chinese border that included nuclear threats and contingency exercises for operations known as Chequerboard and Falcon. War was averted, though not before local Indian commanders consulted with New Delhi on the possibility of Chinese tactical nuclear use and inquired about plans for an Indian response.[46]

Finally, the most famous example of Indian risk taking was the 1971 war with Pakistan, resulting in creation of the independent state of Bangladesh. Taking advantage of a self-inflicted political crisis in East Pakistan, India adroitly maneuvered to isolate Pakistan internationally. It provided significant support to Bangladeshi nationalist forces, including gradual infiltration of Indian Army units into permanent posts inside East Pakistan. It consolidated a defensive alliance with the Soviet Union, acquired large stocks of Soviet weapons and spares, and timed its military operations to coincide with the least favorable period for Chinese intervention. Although Indian forces were already engaged in combat with Pakistani forces inside East Pakistan, most histories trace the beginning of the war to air attacks by Pakistan against western India, which occurred three days before Indian plans committed it to large-scale conventional military operations. Unlike the ponderous operations in the West, the Indian attack on East Pakistan was rapid, dynamic, and overwhelming, as the Indian commanders had adapted to the challenging terrain and enemy mistakes. The result was a decisive victory that bisected Pakistan and created the Bengali nation.[47]

Archetype Three: Domestic Manager

Of the various types of war, the one in which India might be most consistent and culturally driven is counterinsurgency. Indian efforts in this arena have been almost constant over the sixty-five years of independence. The best-known efforts have been largely confined to the far periphery—namely, states in the northeast (ex–North East Frontier Agency), Jammu and Kashmir, and Punjab in the west. However, the current Naxalite insurgency is based on caste, class, and wealth rather than religion or ethnonationalism and is considered the single most pressing threat.[48]

The Indian approach to counterinsurgency is not unique, but it is consistent. It starts with the determination of the political leadership to safeguard the territorial integrity of the country. Internal weakness, as mentioned earlier, was seen as the reason for the British conquest. Indian leaders fear the domino effect: if one province successfully revolts, then others in the vast multiethnic cultural mélange could follow suit. This is why Kashmir remains such a sensitive matter. India is determined to prove it is a secular society and Muslims have nothing to fear as citizens. Accepting a different status for Kashmir might encourage other Muslims to seek independence, autonomy, or closer links with Pakistan.

Indian counterinsurgency resembles a population-centric unconventional warfare doctrine.[49] However, combining strong military action with efforts at political accommodation is perhaps unique. Indian military and paramilitary forces are both heavy-handed and recipients of high casualties. Counterinsurgency operations may last for decades and intensify even during periods of intense political activity. Indian efforts are aimed at dealing maximum damage to violent intransigents while simultaneously encouraging alienated people to reenter the political system. Most efforts end with insurgents integrated into the political system and a gradual winding down of military operations. The wear and tear is substantial on Indian forces, who exhibit extraordinary patience. Although the army is not always pleased to participate in counterinsurgencies and recently has been replaced with dedicated paramilitary units (often recruited from among former soldiers), it has been successful in a number of difficult and bloody campaigns.[50]

A Modern Indian Way of War?

These archetypes represent patterns in the use of the Indian military since independence. It is tempting to impose a cultural framework on these patterns and to imply that they somehow incorporate a common tradition that dates back to Kautilya and *The Mahabharata*. Nevertheless this would be difficult to prove and maintain, especially given the significant discontinuities in Indian culture and military practice from 1526 until Independence. These archetypes combine to form a pattern of modern Indian political culture that may reveal a much older tradition. Caution, patience, stubbornness, and deliberation are qualities of Indian elites. Thus, it is unsurprising that such traits are manifested in risk aversion and domestic management as well as national security policy in general. India has remained relatively secure by virtue of its territory, population, economic resources, and a highly conservative approach to foreign and defense affairs that makes sense and complements the ideology of modern India's founding leaders.

Risk taking, the third archetype, is different and far more interesting. It is quite tempting to accept this standard as another aspect of Indian culture: an emerging entrepreneurial India of the twenty-first century. Again, it probably attributes too much explanatory power to an oversimplified vision of culture. However, important elements are common in each period when this archetype has emerged. The first and most relevant element is that risk taking is associated with new and more dynamic political leadership. These leaders may take risks that distinguish them from their predecessors and capitalize on emerging opportunities. Indira Gandhi became known internationally after 1971. Leaders of the Bharatiya Janata Party used nuclear tests to position themselves on the global and domestic political stage. Events in 1986 were the result of an unusual troika that included Rajiv Gandhi, General Krishnaswamy Sundarji, and Arun Singh, who were determined to test the limits of newly acquired military power.

A second common element was a sense of timing. In each case, an opportunity appeared in conjunction with a sense that the window might close quickly. It was reported that Indira Gandhi initially wanted to go to war in late spring only to be told by her generals that it was impossible. The 1986–1987 crises occurred under the shadow of Pakistan's nuclear proliferation; General Sundarji hoped to exercise India's conventional military advantage before the opportunity was denied by the Pakistani nuclear threat.[51] In 1998, India feared the Comprehensive Test Ban Treaty or some other US-led nonproliferation regime would lock it into a nonnuclear status.

Finally, each case depended on acquiring powerful leverage. In 1971 and 1986, India had achieved periods of rapid and significant military buildup, particularly on the ground. In 1971, Pakistan also was conducting genocidal repression of Bengali nationalists, and then, in 1986, it was distracted by its involvement in supporting Afghan resistance to Soviet occupation. But in 1998,

Indian leverage was somewhat unusual. The economic crisis during the previous year had ravaged Asia, but India suffered relatively little because of its limited integration in the global economy. Consequently, powerful economic growth introduced by reforms in 1990 continued unabated, which gave New Delhi significant global influence. Even when nations in the West placed sanctions on India, they were comparatively mild and did not target the banking sector. India's undisputed standing as a major emerging economy furnished a substantial deterrent to coercion of its financial sector and tempered international responses.

• • •

In cultural terms, an Indian way of war probably does not exist. However, a limited and nuanced approach can identify tendencies in national security planning and policies that can be linked to the political culture, traditions, and preferences of modern Indian leadership. Such tendencies engender patient, methodical, and relatively risk-averse approaches to international and internal conflicts. However, when certain factors align, such as new leaders, international opportunities, or a sense of urgency, and a belief that India has a particular advantage that can prove decisive, Indian risk aversion can be reversed and caution thrown to the winds.

NOTES

1. Among them are George Modelski, "Kautilya: Foreign Policy and International System in the Ancient Hindu World," *American Political Science Review* 58 (September 1964): 549–60; Stephen Peter Rosen, *Societies and Military Power: India and Its Armies* (Ithaca, NY: Cornell University Press, 1996); and George K. Tanham, *Indian Strategic Thought: An Interpretive Essay* (Santa Monica, CA: RAND, 1992).

2. Thomas Cleary, *The Japanese Art of War: Understanding the Culture of Strategy* (Boston: Shambhala Press, 1991); Russell F. Weigley, *The American Way of War: A History of United States Military Strategy and Policy* (Bloomington: Indiana University Press, 1973). On China, see Alastair Iain Johnston, *Cultural Realism: Strategic Culture and Grand Strategy in Chinese History* (Princeton, NJ: Princeton University Press, 1995); Chen-Ya Tien, *Chinese Military Theory: Ancient and Modern* (Oakville, ON: Mosaic Press, 1992); and Arthur Waldron, "Chinese Strategy from the Fourteenth to the Seventeenth Centuries," in *The Making of Strategy: Rulers, States and War*, ed. Williamson Murray, Macgregor Knox, and Alvin Bernstein (Cambridge: Cambridge University Press, 1994), 85–114.

3. Samuel P. Huntington, *The Clash of Civilizations and the Remaking of World Order* (New York: Simon and Schuster, 1996).

4. Rosen, *Military Power*, 1–32; P. C. Chakravarti, *The Art of War in Ancient India* (New Delhi: D. K. Publishers, 1941).

5. The English translation by Kisari Mohan Ganguli (1883–1896) of *The Mahabharata of Krishna-Dwaipayana Vyasa* by Kamala Subramaniam has been published in four volumes (Philadelphia, PA: Coronet Books, 1991). *The Arthashastra* by Kautilya has been translated by R. P. Kangle (Delhi: Motilal Banarsidass, 1972) and L. N. Rangarajan (New Delhi: Penguin Books, 1992).

6. Rosen, *Military Power*, 62.

7. Chakravarti, *Art of War*, iv–v.

8. Niccolo Machiavelli, *The Prince* (New York: Barnes and Noble, 1999); Rosen, *Military Power*, 62.

9. Rosen, *Military Power*, 67.

10. Chakravarti, *Art of War*, 12.

11. Gurcharin Singh Sandhu, *A Military History of Ancient India* (New Delhi: Vision Books, 2000), 328–36.

12. R. Ernest Dupuy and Trevor Dupuy, *The Encyclopedia of Military History from 3500 BC to the Present*, 2nd rev. ed. (New York: Harper and Row 1986), 394–96.

13. Jos Gommans, *Mughal Warfare* (New York: Routledge, 2002).

14. The Greek invasion incorporated the four northwestern kingdoms into the Mauryan Empire in short order. Sandhu, *Military History*, 211.

15. Dupuy and Dupuy, *Encyclopedia*, 194–96; Sandhu, *Military History*, 369–74.

16. Dupuy and Dupuy, *Encyclopedia*, 392–96, 440–41.

17. H. S. Bhatia, *Military History of British India* (New Delhi: Deep and Deep, 1977), 158–68; Anjali Nirmal, *The Decisive Battles of Indian History* (Jaipur: Pointer, 1999), 112–246.

18. Nirmal, *Decisive Battles*, 277–384; Dupuy and Dupuy, *Encyclopedia*, 699–702, 784–89.

19. Bhatia, *Military History*; Randolph G. S. Cooper, *The Anglo-Maratha Campaigns and the Contest for India* (Cambridge: Cambridge University Press, 2003).

20. Peter Ward Fay, *The Forgotten Army: India's Armed Struggle for Independence, 1942–1945* (Ann Arbor: University of Michigan Press, 1993).

21. Chakravarti, *Art of War*, 11–14; Sandhu, *Military History*, 372–74; Gommans, *Mughal Warfare*, 67–97; Rosen, *Military Power*, 104–61.

22. Waheguru Pal Singh Sidhu, "Of Oral Traditions and Ethnocentric Judgments," in *Securing India*, ed. Kanti Bajpai and Amitabh Mattoo, 174–90; Chakravarti, *Art of War*, i; Rosen, *Military Power*, 62–65.

23. Gommans, *Mughal Warfare*.

24. Chakravarti, *Art of War*, cites eighth-century AD sources referring to the use of chariots in war.

25. Ibid., 51, mentions the use of war elephants in a battle in the twelfth century.

26. Ibid., 36–43.

27. Nirmal, *Decisive Battles*, 277–98; Jeremy Black, *War and the World: Military Power and the Fate of Continents, 1450–2000* (New Haven, CT: Yale University Press, 1998), 111–14.

28. On Indian cultures after conquest, see John A. Lynn, "Victories of the Conquered: The Native Character of the Sepoy," in *Battle: A History of Combat and Culture from Ancient Greece to Modern America*, ed. John A. Lynn (Boulder, CO: Westview, 2003), 145–77.

29. Dupuy and Dupuy, *Encyclopedia*, 506.

30. Rosen, *Military Power*, 140–57. As Mughal armies incorporated more Hindu officers, they became less efficient and more torn by internal turmoil—maximum military efficiency was reached in the sixteenth century when the officer corps was non-Hindu.

31. See Chandar S. Sundaram, "Grudging Concessions: The Officer Corps and Its Indianization, 1817–1940," in *A Military History of India and South Asia from the East India Company to the Nuclear Era*, ed. Daniel P. Marston and Chandar S. Sundaram (Westport, CT: Praeger Security International, 2007), 88–101.

32. Douglas M. Peers, "The Martial Races and the Indian Army in the Victorian Era," in Marston and Sundaram, *Military History*, 34–52.

33. Lynn, "Victories of the Conquered."

34. Tanham, *Indian Strategic Thought*; see also Raju G. C. Thomas, *Indian Security Policy* (Princeton, NJ: Princeton University Press, 1986), 86–90, 135–38.

35. Lorne J. Kavic, *India's Quest for Security: Defence Policies, 1947–1965* (Berkeley: University of California, 1967), 141–68.

36. Stephen P. Cohen and Sunil Dasgupta, *Arming Without Aiming: India's Military Modernization* (Washington, DC: Brookings Institution, 2010).

37. George Perkovich, *India's Nuclear Bomb: The Impact on Global Proliferation* (Berkeley: University of California, 1999), 13–22; see also Jaswant Singh, *Defending India* (New York: St. Martin's, 1999), viii–xxvi, 1–60.

38. On the use of militants by Pakistan and its domestic impact, see Hassan Abbas, *Pakistan's Drift into Extremism: Allah, the Army, and America's War on Terror* (New Delhi: Pentagon Press, 2005); Bruce Riedel, *Deadly Embrace: Pakistan, America, and the Future of the Global Jihad* (Washington, DC: Brookings Institution, 2011); and Zahid Hussain, *Frontline Pakistan: The Struggle with Militant Islam* (New York: Columbia University Press, 2007).

39. Weigley, *American Way of War.*

40. Sumit Ganguly, *Conflict Unending: India-Pakistan Tensions Since 1947* (New York: Columbia University Press, 2001); Steven A. Hoffman, *India and the China Crisis* (Berkeley: University of California Press, 1990).

41. Robert G. Wirsing, *Kashmir in the Shadow of War: Regional Rivalries in a Nuclear Age* (London: M. E. Sharpe, 2003); Victoria Schofield, *Kashmir in Conflict: India, Pakistan and the Unending War* (London: I. B. Tauris, 2000); Sumit Ganguly, *The Crisis in Kashmir: Portents of War, Hopes of Peace* (Cambridge: Cambridge University Press, 1997).

42. K. C. Paval, *Indian Army After Independence*, rev. ed. (New Delhi: Lancer, 1990), 167–71; *Freedom to Use the Seas: India's Maritime Military Strategy* (New Delhi: Integrated Headquarters, Ministry of Defence [Navy], 2007).

43. Perkovich, *India's Nuclear Bomb*, 353–444.

44. Raj Chengappa, *Weapons of Peace: The Secret Story of India's Quest to Be a Nuclear Power* (New Delhi: HarperCollins, 2000).

45. Ravi Rikhye, *The War That Never Was* (New Delhi: Chanakya, 1988); Sumit Ganguly and Devin T. Hagerty, *Fearful Symmetry: India-Pakistan Crises in the Shadow of Nuclear Weapons* (New Delhi: Oxford University Press, 2005), 68–81; Kanti P. Bajpai, Pervaiz I. Cheema, S. Ganguly, P. Chari R., and S. P. Cohen, *Brasstacks and Beyond: Perception and Management of Crisis in South Asia* (Urbana-Champaign: Program in Arms Control, Disarmament, and International Security, University of Illinois, 1995).

46. Perkovich, *India's Nuclear Bomb*, 289.

47. Robert Jackson, *South Asian Crisis: India-Pakistan-Bangladesh* (London: Chatto and Windus, 1975); D. K. Palit, *The Lightning Campaign: Indo-Pakistan War, 1971* (New Delhi: Thompson Press, 1972); Richard Sisson and Leo E. Rose, *War and Secession: Pakistan, India, and the Creation of Bangladesh* (Berkeley: University of California Press, 1990).

48. V. N. Singh, *Naxalism: A Great Menace* (New Delhi: Prashant, 2010); Manoj Kr. Singh and S. K. Chaudhary, *Maoist Guerrilla: Seeing Red* (New Delhi: Surendra, 2010).

49. For a recent effort to codify this concept, see United States Department of the Army, *Counterinsurgency*, Field Manual 3-24 (Washington, DC: Headquarters, Department of the Army, 2006).

50. Thomas, *Indian Security Policy*, 51–85; C. Christine Fair, *Urban Battle Fields of South Asia: Lessons Learned from Sri Lanka, India, and Pakistan* (Santa Monica, CA: RAND, 2004).

51. P. R. Chari, Pervaiz Iqbal Cheema, and Stephen P. Cohen, *Four Crises and a Peace Process: American Engagement in South Asia* (New Delhi: HarperCollins, 2008), 67.

10 MILITARY MODERNIZATION IN ASIA

Richard A. Bitzinger

FEW PARTS OF THE WORLD are undergoing higher levels of military modernization than Asia. Fueled by strong economies and matching defense outlays, the region has experienced a veritable binge in arms spending since the 1990s. Even the financial crisis in 1998 and the recession from 2008 to 2010 failed to diminish weapons procurement. Accordingly, over the past two decades many militaries in the region have almost completely recapitalized their force structures. In fact, future acquisitions will generate new defense establishments. This dynamic is significant because the Asia-Pacific region remains prone to war. Aside from the Middle East, few areas are more perilous than the Korean Peninsula, Taiwan Strait, South China Sea, and Sino-Indian border.

RECENT DEVELOPMENTS

As the result of indigenous weapons production and arms imports, Asian militaries are acquiring capabilities for power projection, precision strike, long-range attack, and battlespace intelligence. Such assets include surface combatants, amphibious assault ships, aircraft carriers, submarines, missile defenses, advanced fighters, antiship and land-attack ballistic missiles, standoff weapons, and smart bombs. Moreover, these weapons are complemented by improved command, control, communications, computers, intelligence, surveillance, and reconnaissance (C4ISR) systems; unmanned aerial vehicles; and airborne early-warning systems.

Surface Combatants

Asian navies have expanded considerably over the last two decades in their size and competence. Regional fleets that once were oriented mainly toward coastal defense are being upgraded to operate in littoral and even blue-water (open ocean) environments. Many nations have consequently acquired larger surface combatants for their navies.

The People's Liberation Army Navy (PLAN) has built at least seven destroyers of different types since 2000, including one class (type 052C/D) with Aegis-type air-defense radar and fire-control systems and YJ-83 or YJ-62 antiship cruise missiles (ASCMs). Moreover, they are armed with Chinese-built vertical-launch HHQ-9 surface-to-air missiles (SAMs). Beijing also has acquired at least a dozen type-054A-class frigates (with six additional ships likely to be constructed) with stealthy design and ASCMs and vertical-launch system-deployed SAMs. Other ships include new-generation type 022 *Houbei*-class catamaran-hull missile fast-attack craft with YJ-83 ASCMs, of which at least sixty have been built. Moreover, China purchased four *Sovremenny*-class Russian destroyers with 3M-80E Moskit (SS-N-22 Sunburn) ramjet-powered, supersonic ASCMs with ranges of 120 to 200 kilometers.

Australia is building three air-warfare destroyers armed with US Aegis combat systems and Standard SM-2 SAMs. These *Hobart*-class destroyers support the expeditionary strategy currently being implemented by the Royal Australian Navy. The keel of the first vessel, Her Majesty's Australian ship (HMAS) *Hobart*, was laid down in 2012, to be followed by work on HMAS *Brisbane* and HMAS *Sydney*. These new destroyers significantly augment existing defense capabilities, from area air defense and escort duties to peacetime military activity and diplomatic missions.

South Korea contracted for the construction of three 7,700-ton KDX-III destroyers (known as the *King Sejong the Great* class) with an option for three more. These ships are equipped with US Aegis air-defense radar and fire-control systems and Standard SM-2 Block IIIB air-defense missiles. Like Australian air-warfare destroyers, KDX-IIIs can be upgraded with SM-3 missiles, and they are also armed with Hyunmoo-IIIC land-attack cruise missiles (LACMs) and Harpoon or indigenous Haesung (Sea Star) ASCMs in 128 vertical-launch cells.

The Republic of China (ROC) Navy has almost recapitalized its fleet since 1990 with the purchase of six *Lafayette*-class frigates from France and the production under license of eight US-designed *Perry*-class frigates. Moreover, Taiwan bought four ex-*Kidd*-class destroyers and eight ex-*Knox*-class frigates from the United States. These warships are equipped with Harpoon or indigenously made Hsiung Feng II/III ASCMs. The *Kidd*-class destroyers also are armed with Standard SM-2 missiles for area air defense. In addition, Taiwan is acquiring thirty-one *Kuang Hua* VI fast missile boats. Finally, it is reportedly developing the stealthy *Hsun Hai* (Swift Sea) corvette that will likely be armed with Hsiung Feng III supersonic ASCMs.[1]

The transition from brown water to the open ocean has been pronounced in Southeast Asia. The Republic of Singapore Navy recently bought six *Formidable*-class frigates, which are based on the French *Lafayette*-class stealth vessel and armed with both Harpoon ASCMs and Aster-15 air-defense missiles. Indonesia is acquiring four Dutch *Sigma*-class corvettes with Chinese C-802 (YJ-83) antiship missiles, and Malaysia has built six German-designed MEKO A100 offshore patrol boats and is currently buying two British *Lekiu*-class frigates and a half-dozen littoral combat ships based on the French *Gowind* design.

Force Projection and Expeditionary Warfare

Many navies have expanded force projection and expeditionary warfare capabilities with rotary- and fixed-wing aircraft. China launched type 071 landing platform docks, twenty-thousand-ton amphibious warfare ships with helicopters and air-cushioned landing craft (LCAC) for eight hundred troops. It reportedly may purchase up to eight type 071 vessels to be complemented by new, larger amphibious assault ships of the landing-helicopter-dock (LHD) type.[2]

Perhaps the most dramatic development was the delivery of the first PLAN aircraft carrier *Liaoning*, the rebuilt Soviet carrier *Varyag* that was acquired in 2001 and underwent initial sea trials in 2011. For a long time, China lacked fixed-wing carrier-based aircraft, but this problem was apparently resolved with the acquisition of the J-15 fighter (a reverse-engineered version of the Su-30, apparently purchased from Ukraine). In all likelihood, the carrier will be used for research and training rather than combat, though it could be pressed into limited service. Nevertheless, it was thought at one time that the Chinese might be planning to construct up to six indigenous carriers.

Japan has expanded its power-projection capabilities with high-speed sealift vessels and three large amphibious *Osumi*-class ships with helicopter decks. Known as a landing ship tank, this thirteen-thousand-ton vessel can transport 330 troops and ten tanks, and has four helicopters and two LCAC hovercrafts. Moreover, the Japanese in the late 2000s acquired two fourteen-thousand-ton *Hyuga*-class helicopter destroyers (DDHs), and they are currently building two nineteen-thousand-ton *Izumo*-class DDHs. These ships feature through-deck design and below-deck hangars that resemble small carriers similar to the Spanish *Principe de Asturias*– or British *Invincible*-class carriers, which are deployed with antisubmarine-warfare helicopters.

In 2005 South Korea deployed the first *Dokdo*-class amphibious assault ship, which displaces fourteen thousand tons and can embark seven hundred troops, ten tanks, fifteen helicopters, and two air-cushioned landing craft. This vessel serves as a multifunctional rapid-response fleet command ship for destroyers, frigates, and submarines.[3] Orders for at least one additional *Dokdo*-class ship were placed with Korean shipyards, with the first in class being commissioned in 2007.

The growing Australian requirement for amphibious and expeditionary warfare capabilities includes the transportation and sustainment of three thousand troops. Consequently, the Royal Australian Navy plans to acquire two new twenty-thousand-ton *Canberra*-class amphibious power-projection LHD ships, each capable of deploying one thousand men and 150 vehicles (including M1A1 Abrams tanks), with landing craft and transport and battlefield-support helicopters. These ships are based on a Spanish design and perform air support, amphibious assault, transportation, and command center roles. The program cost is 3 billion Australian dollars, and the first of the LHD ships is scheduled to enter service sometime in 2014.[4]

While Japan, South Korea, and Australia have no plans to buy fixed-wing carriers, their open-flight-deck helicopter ships—the *Hyuga* and *Izumi* classes, *Dokdo*, and *Canberra*—might conceivably be modified with a ski-jump deck to serve as pocket carriers with short take-off/vertical landing (STOVL) combat aircraft such as F-35B Joint Strike Fighters (JSFs). The *Canberra*-class LHD retains the ski jump as part of its original design.

India is investing in a carrier-centered force to replace its aging British-built carriers with the acquisition of the Soviet-built *Admiral Gorshkov*, which is a forty-five-thousand-ton *Kiev*-class carrier decommissioned in 1996. After years of negotiations, Moscow struck a deal that provided the carrier gratis, while New Delhi paid approximately $1 billion to refit and upgrade the vessel to launch MiG-29s with short take-off but arrested recovery (STOBAR).[5] This entailed stripping all weapons from the foredeck and adding a 14.3-degree ski jump and arrestor wires.

Moreover, India will spend another $700 million on aircraft and weapons systems, which include twelve single-seat MiG-29K Fulcrum-D fighters, four dual-seat MiG-29KUB trainers, and six Kamov Ka-27 and Ka-31 helicopters together with simulators, spare parts, training, and maintenance facilities.[6] The carrier, renamed Indian Navel Ship (INS) *Vikramaditya*, was scheduled for delivery in 2008, but its refitting was more difficult than expected, resulting in considerable cost overruns. Moscow asked for an additional $1.2 billion to complete the upgrade. As a result, the *Vikramaditya* was not commissioned until late 2013.[7]

The Indian Navy has similar problems with its indigenous carrier that had been known as an air-defense ship. The 37,500-ton carrier is named INS *Vikrant* and will use both ski jumps and arrestor wires to support the MiG-29K and Indian Tejas light combat aircraft, which will not enter service until 2015 at the earliest.[8] Consequently, India is extending the life of the fifty-year-old *Viraat* (ex-British *Hermes* with Harriers) for four or five years. Ultimately, the Indians want carrier battle groups to operate on both coasts and the *Viraat* in reserve.[9]

Several nations in Southeast Asia have been acquiring ships for expeditionary amphibious warfare. The Republic of Singapore Navy, for example, operates two indigenously designed and constructed *Endurance*-class landing ships, each capable of carrying 350 troops, eighteen tanks, four helicopters, and four landing craft. Meanwhile, both Indonesia and Malaysia are buying or considering buying foreign-built landing platform dock (LPD) warships. Finally, the Royal Thai Navy is the only regional force besides India with a fixed-wing aircraft carrier, the ten-thousand-ton Spanish-built *Chakri Nareubet*. This ship has nine AV-8A Harrier STOVL jets and six S-70B Seahawk helicopters but rarely puts to sea because of the cost and operational reliability of its Harriers.

Submarines

China has greatly expanded its submarine force over the last fifteen years. Since the late 1990s it has acquired thirteen type 039 *Song*-class diesel-electric submarines. The boats are the first domestically built submarine to feature modern teardrop hulls; skewed quiet propellers; encapsulated ASCMs, which can be fired while submerged through regular torpedo tubes; and antisubmarine rockets. The type 41 *Yuan*-class boat that appeared in 2005 has superseded them. The *Yuan* submarines also are armed with torpedoes and ASCMs and perhaps have Stirling air-independent propulsion engines as found in both Japanese and Swedish submarines. Four *Yuan*-class boats have been completed and at least three more are being built.

China is replacing its five *Han*-class nuclear-powered attack submarines (SSNs) and one *Xia*-class nuclear-powered ballistic-missile submarine (SSBN). The first type 093 *Shang*-class SSN was commissioned in 2006.[10] The PLAN has launched two type 094 *Jin*-class SSBNs, with twelve JL-2 submarine-launched ballistic missiles (SLBMs) having a range of seven thousand kilometers—three times that of JL-1 SLBMs on *Xia* boats.[11] Since the mid-1990s, China has bought twelve Russian *Kilo*-class diesel-electric submarines, which are armed with 3M-54E Klub (SS-N-27) ASCMs and 53-65KE wake-homing torpedoes; some *Kilo* features have been allegedly incorporated in *Yuan*-class submarines.

Japan is building *Soryu*-class diesel-electric submarines equipped with Stirling engines for air-independent propulsion. At least four boats in this class are under construction and five more are planned; they will be built at a rate of roughly one submarine a year.

Australia took delivery of six *Collins*-class submarines from 1990 to 2003. These boats can patrol hostile waters virtually undetected to gather intelligence, undertake surveillance, and from above and below the surface conduct reconnaissance as well as other missions on land and in the air. Canberra announced in 2009 that the *Collins*-class submarine would be replaced by 2030 by a dozen new boats to defend sea approaches, support other forces, and conduct operations when stealth and other advanced capabilities are decisive. A larger submarine force will increase the challenges of additional defensive capabilities for an adversary.

South Korea is enlarging its fleet of submarines. During the 1990s it bought nine German-designed type 209 diesel-electric submarines of the KSS-I *Changbogo*-class, which are being replaced by the German type 214 *Chungji*-class (KSS-II). Three type 214 submarines were constructed under license, with the first commissioned in 2008 and options for six more boats. Type 214 submarines have fuel cells for air-independent propulsion and can submerge for up to three weeks.[12] However, Seoul may decide to build its own KSS-III boats, which depending on the number ordered, might include eighteen operational submarines by 2020–2025.[13]

After protracted negotiations, India signed an agreement for six Franco-Spanish *Scorpène*-class submarines to be constructed under license at the Mazagon Docks shipyard; six additional boats may be ordered with the French *Module d'energie sous-marine autonome* system for air-independent propulsion.[14] New Delhi also wants to develop nuclear submarines, has leased two *Akula*-class boats from Russia, and is attempting to build its own class of hunter-attack and ballistic-missile-armed nuclear-powered submarines, based on its advanced technology vehicle (ATV) project. India has worked on the ATV program for over thirty-five years, and it launched the first vessel in this class, the INS *Arihant*, a ballistic-missile-armed nuclear-powered submarine, in 2009. The Indian Navy would like to deploy by 2023 a fleet of up to four SSBNs, which will be armed initially with the indigenously developed Sagarika SLBMs.[15]

Singapore had no submarines until it purchased four 1960s-era boats from Sweden during the 1990s. In 2009 it bought two more boats from Sweden that were renamed *Archer*-class and retrofitted with Stirling engines with air-independent propulsion. In 2013, Singapore announced that it would acquire two brand-new type 218 submarines from Germany.[16] Malaysia has taken delivery of two Franco-Spanish *Scorpène*-class submarines that were commissioned in 2009, and Vietnam is acquiring six Russian *Kilo*-class diesel-electric submarines for $2 billion.[17] In the mid-2000s, Indonesia was rumored to be purchasing four *Kilo*-class and two *Lada*-class submarines to replace German-built type 209 boats. However, the deal fell through when Moscow refused to permit Jakarta to use credits to build a submarine base. Indonesia subsequently agreed to buy three submarines from South Korea (based on the German type 209 design). Jakarta has requirements to acquire up to six boats in all. Thailand also has identified requirements for two or more submarines.

Taipei has long tried to acquire modern diesel-electric submarines to replace its two aging Dutch-built boats. Although Washington had agreed to furnish diesel-electric submarines, no shipyard in the United States builds them. After more than a decade of seeking another source, no nation has been willing incur the wrath of China to construct them.

Fighter Aircraft

Modernization of the PLA Air Force (PLAAF) and PLAN Air Force (PLANAF) has focused on acquiring fighters with advanced air-to-air missiles (AAMs), air-to-ground weapons, and long-range SAM systems. In the last fifteen years, China has acquired large inventories of so-called fourth-generation and fourth-generation-plus fighters with either standoff active-radar-guided medium-range AAMs or precision-guided air-to-surface munitions. In 1992, for example, the Chinese began importing Russian-built Su-27 fighters, which were complemented by the purchase of more-advanced Su-30MKKs. Moscow agreed to produce Su-27s (designated J-11As) under license in Manchuria. All together, China has bought three hundred Su-27s (including 100 J-11As) and Su-30MKKs. For a decade it has manufactured a reverse-engineered version of Su-27s that it designates J-11Bs, albeit with Russian engines.

China also is producing an indigenous fourth-generation-plus combat aircraft. The J-10 is an agile fighter in a class similar to the F-16C with fly-by-wire flight controls and a glass cockpit but equipped with the Russian AL-31 engine, underscoring Chinese difficulties in developing jet engines. The first J-10 flew in the mid-1990s and production began at the turn of the century. Some 150 J-10s have been deployed with about thirty aircraft being delivered each year; estimates are for up to three hundred J-10 fighters to be built. By the end of this decade the PLAAF and PLANAF inventories will have six hundred or more fourth-generation or later combat aircraft.

China has acquired RE-77E (AA-12) active-radar-guided AAMs for its Su-27s, while Su-30s can be equipped with Russian Kh-31P antiradiation missiles to use against radar. The J-10 fighters have Chinese-designed PL-12 active-radar AAMs; short-range PL-8s, which are variants of Israeli Python-3; laser- and satellite-guided bombs; high-speed antiradar missiles; and air-launched cruise missiles.

In a development comparable to launching its first aircraft carrier, in 2011 China unveiled the fifth-generation J-20 fighter. The aircraft resembles the US F-22, although details on its stealth, radar, avionics, and weapons remain sketchy. Consequently, one should not attribute too much to the program.[18] In addition, information has leaked out about a rival Chinese fifth-generation fighter, the J-31, which appears to be a smaller and more agile aircraft, more closely resembling the US F-35 JSF, albeit with two engines. Although details on the Chinese fifth-generation fighter are imprecise, what is known is that it demonstrates a willingness by China to aggressively pursue reaching the vanguard of fighter production.

Japan bought a hundred indigenous F-2 fighters (modified F-16s) to add to its force of more than two hundred F-15s. After attempting to buy F-22 fifth-generation fighters, Tokyo announced in 2011 that it would acquire forty-two F-35 JSFs and possibly more aircraft in the future. South Korea is acquiring sixty-one F-15Ks and planning to purchase as many as sixty fifth-generation fighters, perhaps of indigenous origin. Australia is replacing its aging fleet of F-111s and F/A-18A/Bs with F-35s. Meanwhile, it has bought twenty-four F/A-18F fighters to fill the gap.

India has imported or produced 240 Su-30MKI fighters and may buy as many as fifty more.[19] It also plans to acquire up to 220 indigenous Tejas light combat aircraft. Moreover, a competition was opened some years ago to provide 126 foreign-made fighters to the Indian Air Force, resulting in an intense contest pitting American, European, and Russian firms against each other. In 2012, the French Rafale was finally chosen; the deal requires local production and considerable offsets (i.e., technology transfers and other assistance).

The ROC Air Force has rebuilt its capabilities since the early 1990s. It bought 150 US F-16A/B and 60 French Mirage-2000-5 fighters that were supplied over a decade. Those planes were complemented by 900 MICA (*missile d'interception et de combat aérien*) AAMs; airborne interceptor missiles (AIMs), including the 480 Magic-2, 600 AIM-7M Sparrow, 1,000-plus AIM-9L/M Sidewinder, and 300-plus AIM-120 advanced medium-range air-to-air missiles (AMRAAMs); and AGM-65 Maverick air-to-surface precision-guided missiles.[20] Under an agreement reached with Taiwan, the United States will be upgrading F-16s to F-16C/D Block 52 Plus standard fighters with active electronically scanned array radar and providing GPS-guided Joint Direct Attack Munition (JDAM) bombs. However, the Obama administration has disapproved the sale of 66 F-16C fighters.

The Republic of Singapore Air Force is the most advanced in the region. For example, it operates 74 F-16s of the latest Block 50/52 Plus. In addition, Singapore placed its first order in 2005 for 24 F-15SG fighter aircraft, the first 12 of which have been delivered and stationed in the United States for training purposes. Singapore also is a partner in the international JSF program and could buy up to 100 F-35 fighters.[21] Indonesia signed a deal with Russia to buy two Su-27s and two Su-30s in 2003 and acquired six Sukhois in 2009. The Indonesian Air Force also had planned at one time to buy up to forty more Sukhois.[22] However, it appears that Jakarta will be purchasing and upgrading up to thirty used US Air Force F-16C/Ds at a cost of $400 million to $600 million to equip two new squadrons.[23] Other recent purchases of fighters in Southeast Asia include eighteen Su-30MKM Flankers by Malaysia, which has plans to buy another eighteen aircraft (either the Swedish Gripen or more Su-30s), and twelve Gripens by Thailand.[24]

Missile Defenses

Australia, China, India, Japan, and Taiwan are acquiring missile defenses. Japan, for example, has recently completed upgrading its six Aegis-type destroyers to the US Navy Sea-based Midcourse Defense (SMD) missile defense mode. This upgrade entails improvements to SPY-1 multifunction phased-array radar and fire-control systems to increase the altitude and range of search, detection, track, and control functions to conduct exoatmospheric antimissile engagements. The program will deploy the Standard SM-3 Block IA missile, which has a third stage for extended range and a Lightweight Exo-Atmospheric Projectile, a kinetic warhead for terminal homing and interception. The Japanese SMD system should be fielded by 2011 and complemented by the land-based Patriot PAC-3 system, which provides endoatmospheric protection against missile threats to the homeland.

Other Asia-Pacific nations are following suit with missile defense plans. India has bought the Israeli Green Pine ballistic-missile early-warning radar and is building a national missile defense system with Russian S-300 SAMs and indigenous exoatmospheric and point-defense missile systems. China conducted a missile defense test in early 2010, while Taiwan is attempting to modify its indigenous Tien Kung II SAM into a working missile interceptor.[25]

The acquisition by Australia and South Korea of Aegis-equipped warships could support national missile defenses using the SMD concept. In particular, Australia is integrating its own Jindalee Over-the-Horizon Radar Network in the US missile defense configuration. Upgrading this long-range aircraft-detection system will enable identification of incoming missiles during their early boost phase.[26] South Korea has announced plans to inaugurate indigenous missile defense by 2012 to defend against the North Korean ballistic-missile threat. This program is likely to include sea-based inceptors at a cost of least $214 million.[27] Taiwan has acquired Patriot PAC-3 missiles, which are being complemented by an indigenous missile defense system based on the domestic Skybow III missile.

Long-Range, Precision-Strike Weapons

At least as important as the acquisition of modern platforms throughout the Asia-Pacific region is the steady proliferation of precision-guided weapons for standoff strike. As noted previously, many surface combatants and submarines are being deployed with advanced ASCMs, such as the Harpoon on *Hobart*-class destroyers by Australia, *Formidable*-class frigates by Singapore, *Soryu*-class submarines by Japan; the Exocet on *Scorpène*-class submarines by Malaysia and India; and the Russian 3M-80E Moskit on *Sovremenny*-class destroyers by China. In cooperation with Russia, India has developed BrahMos supersonic ASCMs, which will be fielded in sea-, land-, and air-based modes, and Taiwan is developing the Hsiung Feng III supersonic ASCM.

Many nations are acquiring active-radar-guided, medium-range AAMs for their fighters. These include US-produced AMRAAMs for Australia, Japan, South Korea, Singapore, and Thailand; Russian R-77/AA-12s for China, Indonesia, and Malaysia; and PL-12s for China. Until recently AMRAAMs were embargoed for sale to many nations in the region.

Asia-Pacific armed forces are being increasingly equipped with standoff land-attack munitions. Japan, South Korea, and Singapore are buying GPS-guided JDAMs, and Australia and Singapore are buying the Joint Stand-Off Weapon (JSOW), a precision-guided glide bomb with a range of up to 130 kilometers. Importantly, several nations in the region have developed LACMs, many adapted from existing ASCMs. Taiwan, for example, is deploying the Hsiung Feng IIE (HF-2E) LACM, based on its HF-2 antiship missile. China has developed the Dong-Hai 10 (DH-10) LACM, and South Korea the Hyunmoo-IIIC LACM.

Finally, it is important to take into account the strike value of ballistic missiles armed with nonnuclear warheads. China has deployed conventionally armed surface-to-surface missiles that include the three-hundred-kilometer-range DF-11 (CSS-7) and six-hundred-kilometer-range DF-15 (CSS-6) short-range missiles as well as an arsenal of DF-31 (CSS-9) road-mobile, solid-fuel intercontinental ballistic missiles (ICBMs) with a range of eight thousand kilometers and submarine-launched JL-2s (CSS-N-4) missiles. Finally, DF-21Ds (Dongfeng) antiship ballistic missiles (ASBMs), which are land-based version of JL-1 submarine-launched ballistic missiles, were reportedly deployed in 2010 and have an estimated range of up to 2,700 kilometers.

Moreover, India has developed short-range Prithvi and medium-range Agni missiles and field-tested an SLBM. Other missiles deployed in the region include the MGM-140 Army Tactical Missile System (ATACMS) in South Korea, the US-built HIMARS multiple rocket launcher in Singapore, and the Brazilian ASTROS-II artillery rocket in Malaysia.

C4ISR Capabilities

Finally, many Asian militaries are expanding and upgrading their C4ISR assets.[28] For example, China, Japan, Singapore, and Taiwan have airborne early-warning and command aircraft, while Australia, India, South Korea, and Thailand intend to buy such aircraft in the near future. Japan and South

Korea have Aegis sensor and combat systems deployed on their largest surface ships, and Taiwan is acquiring long-range early-warning radar. Some regional powers are purchasing unmanned aerial vehicles and launching satellites for surveillance, communications, navigation, and target acquisition. All these nations exploit imagery from commercial earth-observation satellites such as Ikonos and QuickBird and the Landsat systems operated by the US Earth Resources Observation and Science Center.

Many Asian militaries are investing in information processing, command and control, and communications and data links. Australia is enhancing firepower, survivability, interoperability, and network-enabled capabilities.[29] South Korea is acquiring integrated tactical communication systems, and Taiwan is spending billions of dollars on C4ISR to link its sensors, computers, and communications among the services.[30] Singapore has gotten C4I network capabilities with fiber-optic and microwave channels to integrate air and maritime surveillance. Also, it is emphasizing the expansion of significant capabilities for network-centric warfare.[31]

China has been improving C4ISR assets on the basis of its conceptions of information warfare. The results are communication, surveillance, and navigation satellites; a manned space program for military use; and experiments with digitized land warfare similar to efforts by the US Army. Additionally, China has invested resources in a separate military communications network with fiber-optic cable, satellites, microwave relays, and long-range high-frequency radio. China also is focused on developing capacities to integrate network warfare, including electronic defenses and countermeasures, computer networks to disrupt enemy computers, and physical attacks on enemy C4SIR networks such as those used as antisatellite weapons.[32]

ARMS IMPORTS

The Asia-Pacific region has long been a major market for advanced conventional weaponry. From 2008 to 2011 it accounted for 42 percent of arms deliveries and 29 percent of arms sales *agreements*.[33] The region was listed second only to the Middle East in arms imports and contained the largest market for weaponry in the world. India, for example, bought almost $21 billion worth of arms from 2008 to 2011, while Taiwan signed deals for $6.5 billion in arms during this period and South Korea for $5.4 billion.[34] Singapore and Malaysia ranked among the largest recipients of weapons systems in Southeast Asia, with Singapore receiving $3.8 billion worth in 2005–2009, becoming the seventh largest in the region.[35] Meanwhile, Kuala Lumpur has acquired more than $5 billion worth of arms since the 1990s.[36] India bought Su-30MKI fighter jets, *Kilo*-class submarines, and T-72 tanks from Russia. Australia, Japan, and South Korea have made major buys of US F-15, F-16, and F/A-18 combat aircraft and naval systems like Aegis air-defense systems and Standard SAMs.

Rising defense budgets reflect much of the growing appetite for arms in the Asia-Pacific region. China, for example, has experienced real double-digit defense increases after inflation almost every year since 1997. Chinese arms expenditures totaled $115 billion by 2013, an increase of 10.7 percent over 2012. Beijing now outspends all other capitals on defense except for Washington. Military spending by China has more than quintupled since the late 1990s, which has enabled China to accelerate military modernization.

Other Asia-Pacific nations have also greatly increased defense expenditures over the past decade. Indian defense spending grew by more than two-thirds from 1998 to 2010, according to the Stockholm International Peace Research Institute,[37] and in 2010 Indian military expenditures totaled roughly $32 billion. Meanwhile, Australia increased defense spending by 46 percent and South Korea by 48 percent over the same period.[38] Of large nations in the region, only Japan and Taiwan maintained relatively fixed levels of spending. In 2008, Taiwan completed a $6.5 billion deal to acquire several US weapons systems that included Patriot PAC-3 missiles, AH-64 attack helicopters, Harpoon ASCMs, and Javelin antitank guided missiles.

According to the Stockholm International Peace Research Institute, the Indonesian military budget trebled from $2 billion in 2001 to $6 billion in 2010, measured in constant 2009 dollars. Malaysian defense spending went from $2.2 billion in 2001 to $3.3 billion in 2010, which represented an increase of 30 percent after peaking at $4.2 billion in 2008. In the same period Thai military spending grew by 58 percent, from $2.7 billion in 2001 to $4.3 billion in 2010. Singaporean defense expenditures rose 28 percent, and the Vietnamese military budget grew 76 percent from 2003 to 2010. Only Taiwan experienced modest growth in defense outlays, amounting to an overall rise of 5 percent over the last decade.[39]

As military procurement by Western nations waned after the Cold War, defense industries went abroad to find new customers to compensate for dwindling domestic markets. In particular, the European, Russian, and Israeli defense sectors now export the majority of their new products. For example,

British-based BAE Systems does less than 25 percent of its business in Great Britain, and French-based Thales derived roughly three-quarters of revenues from outside France. At Swedish-based Saab only around 30 percent of its business is conducted within Sweden. Israeli defense firms export more than 75 percent of products, and Russian arms makers reportedly depend on overseas sales for up to 90 percent of their income.[40] The US defense industrial base has had to rely on exports, even given a huge domestic market, and some products like F-15 and F-16 fighters are now exclusively produced for foreign sales, which is essential to keeping this sector in the black.

Unsurprisingly, large arms-producing nations have come to regard the Asia-Pacific region, which is increasingly bent on acquiring the most sophisticated weapons, as particularly lucrative. For example, from 2004 to 2011 more than $27.5 billion (approximately 60 percent) of arms produced in Russian factories were delivered to the region.[41] While the bulk of its exports once went to China and India, Russia has expanded sales in Southeast Asia, with fighters acquired by Malaysia and Indonesia, submarines by Vietnam, helicopters by Thailand, and tanks by Burma.

The leading Western European arms producers—the United Kingdom, France, Germany, and Italy—also depend heavily on the Asia-Pacific market. Together, these four nations exported approximately $11.4 billion worth of arms to Asia-Pacific nations from 2004 to 2011. The region remains essential for European arms manufacturers.[42] Roughly 41 percent of all French arms exports went to Asia from 2008 to 2011; for Germany, it was 74 percent, and for Italy, 38 percent. Only Britain had a relatively low level of arms exports to the region (33 percent) because of a few good-sized sales in the Middle East, for example, the £5 billion deal with Saudi Arabia for seventy-two Eurofighter Typhoon aircraft.[43]

Between 2004 and 2011, the United States delivered $19.4 billion worth of weapons to the Asia-Pacific region, which is roughly a third of all US arms exports during this period. Only the Middle East, at $42.2 billion, constituted a larger US arms market.[44]

As the competition heated up, supplier restraint was replaced by a readiness on the part of major arms producers to sell just about any type of conventional weapon available. Supplying nations are increasingly prepared to offer technology transfers, licensed production, and other offsets. Germany, for example, has transferred submarine technology to South Korea, and Russia has licensed production of Su-27s to China and Su-30s to India.

LOCAL ARMS INDUSTRIES

While the Asia-Pacific region is a major arms importer, it has also striven to supplement (and in some cases replace) foreign-sourced weaponry with homegrown armaments. Nearly every major nation in the region produces weapons; some have created extensive local arms industries. Some nations are producing arms that approach the state of the art in particular industrial sectors. For example, South Korea manufactures an advanced trainer jet (T-50), and its XK-2 tank is likely as capable as any comparable system in the West. Singapore makes artillery systems and exports arms and armored vehicles. The Indian Tejas fighter uses carbon fiber composites for 45 percent of the airframe. Unsurprisingly, Japan manufactures advanced weapons systems, particularly main battle tanks, submarines, fighter aircraft, and increasingly, missile systems.

Armaments production in the region follows two courses. In China, India, Singapore, and Taiwan, manufacturing is concentrated in government-owned and government-operated factories. China has become self-reliant in arms production and has the largest Asian defense industrial base, with more than a thousand state-owned enterprises and three million workers including more than three hundred thousand engineers and technicians. China is one of few nations to produce military hardware ranging from small arms and armored vehicles to fighters, warships, nuclear weapons, and ballistic missiles. Its systems include J-10 fighters, *Yuan*-class diesel-electric submarines, type 052C destroyers, and HQ-9 long-range SAMs (similar to the Patriot missile). Moreover, China built the world's first ASBM and tested two types of fifth-generation aircraft, while overall production by its military-industrial complex has grown.

India, like China, is an aspiring power that has long harbored the goal of possessing a high-tech, self-sufficient arms industry. These ambitions go back to its attempts in the early 1960s to design and build HF-24 Marut fighters. To this end, New Delhi created a huge government-run sector with state-owned public undertakings, ordnance factories, and research and development. The defense base employs more than 1.4 million workers, including 30,000 in research and development, and enjoys sales exceeding $4 billion per year. Indigenous products include Tejas combat aircraft, Arjun tanks, and BrahMos ASCMs.

Despite having capitalist systems, Taiwan and Singapore rely on state-owned companies to produce arms. Taiwan has developed a family of air-to-air, surface-to-air, and antiship missiles. More recently, it developed LACMs and several kinds of combat aircraft for the air force, including indigenously produced Ching-kuo fighters.

Singaporean arms production is based in the state-owned Singapore Technologies Engineering (STEngg), which makes Bionix fighting vehicles, Terrex personnel carriers, Bronco tracked carriers, Primus and Pegasus 155 mm artillery, and SAR 21 assault rifles. In addition, ST Aerospace (STAe) maintains, repairs, overhauls, and upgrades military and commercial aircraft; for example, it modernized both F-5s and A-4s. STAe is also a member of the US-led consortium developing F-35s and partner to Eurocopter in the production of EC-120 light utility helicopters. Finally, ST Marine has constructed *Fearless*-class offshore patrol vessels, *Endurance*-class amphibious assault ships, and in the last decade, six French-designed *Lafayette*-class frigates designated the *Formidable*-class. STEngg also has considerable expertise in logistics and depot management, maintenance and overhaul of aircraft engines, and ship repair.

On the other hand, Japan and South Korea rely on large conglomerates (*keiretsu* in Japan and *chaebols* in Korea). Since Japan began rearming in the 1950s, it has pursued *kokusanka*, or self-reliance, in weapons manufacturing. Tokyo has put considerable resources into building up and maintaining a technologically advanced defense industrial base with indigenous production. This national policy has been successful; Japan builds tanks, armored vehicles, warships, submarines, and missiles. Japanese Self-Defense Forces depend almost entirely on domestic sources. Moreover, when it imports, because the cost of local development is high, Tokyo secures licenses to manufacture weapons systems domestically.

Arms production in Japan has traditionally involved partnerships between government and industry. Nearly the entire defense sector is based in large, highly diversified companies, such as Mitsubishi and Kawasaki. These firms have received near guarantees to a slice of the annual defense procurement budget, in which each holds a monopoly stake. Mitsubishi, for example, is the only maker of fighters, Kawasaki of large airframes such as the P-1 maritime patrol aircraft and the C-2 transport plane, and Fuji Heavy Industries of turboprop trainers. When two or more contractors build submarines, they function in a duopoly, alternating production contracts. In one instance, two firms merged their shipbuilding businesses into a single company.

South Korea also depends on the private sector, particularly large *chaebols* (industrial conglomerates) such as Samsung, Daewoo, and Hyundai. Defense manufacturing is concentrated in a few *chaebols*, with the seven largest defense contractors accounting for nearly two-thirds of all local procurement. At the same time, the government has been involved in arms production, providing direct and indirect subsidies to industry, underwriting research and development, and designating firms as monopolistic sources of critical military hardware.[45]

The Republic of Korea has developed one of the most impressive defense industrial bases in the Asia-Pacific region. It is particularly broad based, being fueled by extensive investments in aerospace, land systems, and shipbuilding. Nearly 80 percent of South Korean arms are bought domestically, including aircraft, tanks, armored vehicles, warships, and submarines, and it has become increasingly self-reliant in producing missile systems.[46] At first most activity involved the licensed production of foreign military systems, such as F-5 and F-16 fighters and German type 209 submarines. Production gradually became indigenous, including the T-50 supersonic fighters, K-1 main battle tanks, and a series of experimental destroyers. In addition, Korea recently developed ASCMs and LACMs, and it plans to build its own class of attack submarines.

Defense industries in the Asia-Pacific region have made remarkable technological progress over the past two decades. But many remain largely metal-bashers as opposed to true innovators. Although weapons produced in the Asia-Pacific region are mainly good, they are industrial-age as opposed to information-age systems: armor, artillery tubes, surface combatants, and aircraft. Regional firms produce some interesting and even cutting-edge systems like ASCMs and LACMs in South Korea and surveillance satellites in Japan. But the local sources are lacking when it comes to disruptive-type technologies such as network-centric systems and precision-strike weapons. More specifically, they more often than not are deficient in advanced systems integration capabilities that link complex, combined systems of systems.

Overall, most regional defense industrial bases, even in Japan, lack the necessary skills, resources, and technological expertise to be truly innovative. Indigenous defense research and development is small and scarcely comprehensive. Moreover, they also lack the capacity of the US defense sector with its Pentagon budget for research and development that runs to tens of billions of dollars annually. In addition, it operates in a highly inefficient arena of small-scale defense contractors. Just as significantly, the propensity to support costly domestic production for the sake of nationalism or source of supply hobbles regional industries with wasteful and duplicative projects that are uncompetitive with the international market.

The effect of such misguided technonationalism is already evident. The Japanese defense sector has experienced two decades of neglect and is finding production increasingly difficult at traditional levels of *kokusanka*, or autarky; it is doubtful a recent decision to permit limited arms exports will turn this situation around. South Korea may offer the perfect example of technology overreach, as indigenous arms production has bred greater ambition, which fueled the pursuit of programs that reach beyond the economic or technological capacity of the nation.

India is a particularly disheartening case. After China, it has the largest and most ambitious defense industrial base in the Asia-Pacific region, yet its performance over the past fifty years has been weak. Billions of dollars have been squandered on domestic weapons programs that never met requirements and objectives in terms of cost and timeliness. Its defense industries are huge, monopolistic state-owned corporations with a bloated bureaucracy that dominates procurement and seeks indigenous solutions with little attention to capabilities. Despite efforts at reform, the Indian defense industrial base has eluded any effective restructuring.

China is the wild card among regional arms producers. It has faced considerable obstacles in building a defense industrial base. Since the late 1990s the progress of the defense sector has been conspicuous. National science and technology efforts and research and development have been considerably improved, as evidenced by fighter aircraft, warships, submarines, and missile systems. The Chinese defense sector is increasingly gaining on Asian peer competitors, although it probably still lags behind the United States and Western Europe in areas such as jet engines and electronics. However, given the priority that Beijing assigns to military technology, the preferential treatment indigenous arms industries receive, and the high levels of defense spending, further improvements and developments should be expected.

Many nations in the Asia-Pacific region are transforming their militaries, and after at least two decades, the process of modernization is not abating. Instead, Australia, China, India, Japan, Singapore, South Korea, and Taiwan are continuing to acquire new capabilities. Defense outlays are expected to rise, particularly because the recent economic crisis has hardly slowed spending in most Asian nations. Leading suppliers will continue to rely on exports as a way of coping with excess capacity and surplus arms, which means that more-sophisticated weapons will appear on the international market at competitive prices.

The impact of such developments on regional security is unclear but ominous. In the first place, the Asia-Pacific arms buildup has been more than mere modernization; instead, the type of weapons being acquired will significantly upgrade how wars are fought. Regional militaries are fielding systems with greater lethality and accuracy, enhanced battlefield knowledge, better command and control, and improved operational maneuver. Moreover, standoff precision-guided weaponry has boosted effectiveness. Modern submarines and surface combatants, amphibious assault ships, air-refueled combat aircraft, and transport aircraft have extended the range of military actions. Advanced reconnaissance and surveillance platforms have expanded the ability to look over the horizon and in all dimensions. In addition, increased stealth and active defenses such as missile defense and long-range AAMs offer survivability and operational capabilities to regional militaries. Thus, should a conflict arise, it will be fought with greater precision, speed, and range and prove more devastating.

Many Asian militaries, including China, Japan, South Korea, and Singapore, are acquiring weapons that could alter the conduct of war. In particular, those systems with precision-strike, stealth, and C4ISR capabilities provide key ingredients for a revolution in military affairs. Combined, they can develop synergies that create new core competencies. These emerging capabilities, in turn, could potentially affect military operations and alter the nature of war.

The acquisition of new military capabilities has several possible repercussions. From the perspective of a superpower, the purchase of more-advanced weapons by US allies and friends could further regional security by strengthening bilateral alliances, interoperability, and burden sharing. For example, in the last decade, close US allies—Australia, Japan, and South Korea—have imported billions of dollars in arms to modernize their forces. Such enhanced capabilities are crucial to the United States in developing information technologies that could enable allied participation in network-centric warfare. For example, Australia, Japan, and South Korea are acquiring the Aegis naval sensor and combat system, which could permit their ships to link up with US naval forces in cooperative engagements against opposing forces.

The spread of advanced weapons systems could adversely affect regional security when tensions are running high, such as in the Taiwan Strait. The growing Chinese arsenal of ships, aircraft, and precision-guided munitions has intensified perceptions of a cross-strait threat and fueled the acquisition by Taiwan of new air and missile defenses; antisubmarine, antisurface warfare systems; and counterlanding weapons. Yet as military forces become more capable, tension in the Taiwan Strait has not been reduced. Such concern is only multiplied when one considers the types of systems being acquired: transformational weapons that bring fundamental changes in the destructiveness of conventional warfare.

Moreover, without necessarily creating an arms race, military modernization could lead to expensive and ultimately imprudent competition. Although they are defined as noncataclysmic, status quo rivalries that maintain the military balance can be disruptive to regional security and spark arms races. In particular, purchasing advanced weaponry might create security dilemmas that undermine regional stability. Arms procurement by one state, even without an intention to threaten its neighbors, contributes to uncertainty. Reciprocal actions only further raise tensions. Even if the tit-for-tat acquisition of advanced weapons systems does not lead to conflict, it can reinforce mutual suspicions and ultimately have deleterious effects.[47]

Even if conflict can be prevented, it will remain difficult to forestall an arms race or arms competition in the Asia-Pacific region. Weapons acquisition can possess a dynamic of its own, especially when military organizations do not exert self-control in procurement. Nevertheless, security dilemmas caused by the reciprocal arms purchases are human-driven phenomena and accordingly may be controlled by nations with the will to resist them.

• • •

The modernization of armed forces in the Asia-Pacific region has intensified and become a fact of life. In the process China and India have injected uncertainty into the regional security environment through advances in power projection and lethal firepower. Given the proliferation of weapons, future disputes may result in widespread instability and greater destruction.

NOTES

1. "Taiwan to Build New Stealth Warship," *DefenceTalk*, April 19, 2011, http://www.defencetalk.com/taiwan-to-build-new-stealth-warship-33586; Wendell Minnick, "Taiwan Plans Stealthy 900-ton Warships," *Defense News*, April 18, 2010.

2. Ronald O'Rourke, "PLAN Force Structure: Submarines, Ships, and Aircraft," in *The Chinese Navy: Expanding Capabilities, Evolving Roles*, ed. Phillip C. Saunders, Christopher D. Yung, Michael Swaine, and Andrew Nien-Dzu Yang (Washington, DC: National Defense University Press, 2011).

3. "LP-X Dokdo (Landing Platform Experimental) Amphibious Ship," GlobalSecurity.org, November 7, 2011, http://www.globalsecurity.org/military/world/rok/lp-x.htm.

4. Office of the Minister of Defence, "$3 Billion Amphibious Ship Will Strengthen ADF, Boost Australian Industry," June 20, 2007, http://www.defence.gov.au/minister/49tpl.cfm?CurrentId=6780.

5. Rahul Bedi, "Getting in Step: India Country Briefing," *Jane's Defence Weekly*, February 6, 2008.

6. "The Vikramaditya [ex-Gorshkov] Aircraft Carrier," GlobalSecurity.org, September 7, 2011, http://www.globalsecurity.org/military/world/india/r-vikramaditya.htm.

7. Edward Hooten, "Modernizing Asia's Navies," *Asian Military Review* 18 (January 2008): 18; Bedi, "Getting in Step."

8. Hooten, "Modernizing Asia's Navies," 8; "The Vikrant-Class Indigenous Aircraft Carrier (IAC)," GlobalSecurity.org, December 8, 2013, http://www.globalsecurity.org/military/world/india/r-vikrant-2.htm.

9. Hooten, "Modernizing Asia's Navies," 8; "The Vikrant-Class Indigenous Aircraft Carrier (IAC)."

10. "Type 093 (09-III) *Shang* Class Nuclear-Powered Attack Submarine," *SinoDefence*, April 4, 2009.

11. O'Rourke, "PLAN Force Structure," 4–9, 13–18; "Type 094 (09-IV) *Jin* Class Nuclear-Powered Missile Submarine," *SinoDefence*, March 13, 2009.

12. Tim Fish, "Seoul Commissions Type 214 Sub," *Jane's Defence Weekly*, January 23, 2008.

13. Robert Karniol, "Team Prepares for 2007 Start on KSS-III Design," *Jane's Defence Weekly*, December 20, 2006.

14. Rajat Pandit, "India Plans to Buy 6 New Subs, Says Navy Chief," *Times of India*, December 2, 2007.

15. Sandeep Unnithan, "The Secret Undersea Weapon," *India Today*, January 17, 2008.

16. Tim Fish and Richard Scott, "Archer Launch Marks Next Step for Singapore's Submarine Force," *Jane's Defence Weekly*, June 18, 2009.

17. Nga Pham, "Vietnam to Buy Russian Submarines," BBC News, December 16, 2009.

18. John Reed, "J-20 vs. F-35, One Analyst's Perspective," *Defense Tech*, December 31, 2010, http://defensetech.org/2010/12/31/j-20-vs-f-35-one-analysts-perspective.

19. Rajat Pandit, "IAF Wants 50 More Sukhois to Counter China, Pakistan," *Times of India*, October 2, 2009.

20. Stockholm International Peace Research Institute, SIPRI Arms Transfers Database, http://www.sipri.org/databases/armstransfers.

21. Jermyn Chow, "F-15 Training Cements Ties with US," *Straits Times*, November 21, 2009.

22. Trefor Moss, "Painful Progress: Indonesia Country Briefing," *Jane's Defence Weekly*, October 16, 2009; Andrew Tan, *Force Modernization Trends in Southeast Asia* (Singapore: Institute of Defence and Strategic Studies, 2004), 17.

23. John Mcbeth, "'Second-hand' Boost for RI's Air Defense," *Jakarta Post*, October 5, 2011.

24. Tan, *Force Modernization Trends in Southeast Asia*, 17.

25. "China: Missile Defense System Test Successful," *USA Today*, January 11, 2010; "Taiwan to Upgrade to Tien Kung-2 SAM," *Missile Threat*, July 31, 2006.

26. Richard A. Bitzinger, "Asia-Pacific Missile Defense Cooperation and the United States, 2004–2005: A Mixed Bag," Asia-Pacific Center for Security Studies, February 2005, p. 4.

27. "South Korea to Complete Missile Defense by 2012," *Defense News*, February 15, 2010.

28. Jason Sherman, "Digital Drive: Focus, Funding Shifts to C4ISR, Precision Weaponry," *Defense News*, February 16, 2004, pp. 23–24.

29. Australia Government Department of Defence, "Australia's National Security: A Defence Update, 2005," 2005, http://australianpolitics.com/downloads/issues/defence/05-12-15_defence-update-2005.pdf.

30. Sherman, "Digital Drive"; Jason Sherman, "Taiwan to Build Military-Wide C4ISR Network," *Defense News*, October 7, 2003.

31. Bernard Fook Weng Loo, "Transforming the Singapore Armed Forces: Problems and Prospects" (paper presented at the conference "Defense Transformation in the Asia-Pacific: Meeting the Challenge," Honolulu, HI, March 30–April 1, 2004), 5; Tim Huxley, "Singapore and Military Transformation" (paper presented at the conference "The RMA for Small States: Theory and Application," Singapore, February 25–26, 2004), 2.

32. United States Department of Defense, *Annual Report on the Military Power of the People's Republic of China, 2009* (Washington, DC: Office of the Secretary of Defense, 2009), 25–28; You Ji, "China's Emerging National Defense Strategy," Association for Asian Research, January 12, 2005, http://www.asianresearch.org/articles/2428.html; Wendell Minnick, "China Shifts Spending Focus to Info War," *Defense News*, September 11, 2006; Bill Gertz, "Inside the Ring: China Info Warfare," *Washington Times*, June 2, 2010.

33. Richard F. Grimmett and Paul K. Kerr, "Conventional Arms Transfers to Developing Nations, 2004–2011," Congressional Research Service, Report R42678, August 24, 2012, pp. 34, 53.

34. Grimmett and Kerr, "Conventional Arms Transfers," 46.

35. Stockholm International Peace Research Institute, SIPRI Arms Transfers Database.

36. Richard F. Grimmett, "Conventional Arms Transfers to Developing Nations, 2001–2008," Congressional Research Service, Report R40795, September 4, 2009, pp. 34, 37, 45, 48, 51, 59–60.

37. Stockholm International Peace Research Institute, SIPRI Military Expenditure Database, http://milexdata.sipri.org.

38. Ibid.

39. Ibid.

40. Richard A. Bitzinger, "Introduction," in *The Modern Defense Industry: Political, Economic, and Technological Issues*, ed. Richard A. Bitzinger (Santa Barbara, CA: ABC-CLIO, 2009), 5.

41. Grimmett and Kerr, "Conventional Arms Transfers," 52–53.

42. Ibid.

43. David Roberson, "BAE Confirms £5bn Eurofighter Sale to Saudi Arabia," *Times* (London), August 19, 2006.

44. Grimmett and Kerr, "Conventional Arms Transfers," 52–53.

45. Dean Cheng and Michael Chinworth, "The Teeth of the Little Tigers: Offsets, Defense Production, and Economic Development in South Korea and Taiwan," in *The Economics of Offsets: Defense Procurement and Countertrade*, ed. Stephen Martin (London: Harwood), 249; Robert Karniol, "South Korean Industry: Learning Curve," *Jane's Defence Weekly*, October 22, 2003.

46. Jong Chul Choi, "South Korea," in *Arms Procurement Decision Making*, vol. 1, *China, India, Israel, Japan, South Korea and Thailand*, ed. Ravinder Pal Singh (Oxford: Oxford University Press, 1998), 185.

47. Barry Buzan and Eric Herring, *The Arms Dynamic in World Politics* (Boulder, CO: Lynne Rienner, 1998).

11 THE ECONOMIC CONTEXT OF STRATEGIC COMPETITION

Bradford A. Lee

THE MAIN PREMISE of this chapter is that a high probability of strategic competition exists in the Asia-Pacific region in the twenty-first century, above all between China and the United States but also between China and other Asian nations. Chinese words and deeds have displayed classic Thucydidean sources of a rising power's assertiveness: fear, honor, and interest.[1] A second premise is that a major conflict remains improbable. The danger of nuclear escalation would make war very risky, and the extent of economic interdependence would make it costly. Nevertheless, war is possible, because there are so many flash points, and thus opportunities for miscalculation, around the periphery of China.

In view of such probabilities and possibilities, this chapter sheds light on the connections between economic and strategic developments. There are two ways to approach that relationship. One way is to consider the use of economic instruments by those engaged in competition and conflicts. The other is to assess the economic context of probable strategic competition and possible war. The latter is the approach followed in this chapter.

Obvious connections run from economic to strategic developments. The impressive growth of the Chinese economy since the 1980s has facilitated advances in Chinese military capabilities since the 1990s that, in turn, have increased Chinese strategic leverage. Moreover, apparent collisions occur between economic developments and strategic circumstances. Many nations located in the Asia-Pacific region depend on the United States for security but have come to rely on China for their prosperity. How they weigh the relative significance of security considerations against economic considerations will affect whether Sino-American strategic competition favors China or the United States.[2] Two other connections between economics and strategic competition deserve attention: the links between economic crises and strategic competition and the links between long-term potential productivity and ultimate strategic success. Understanding them requires extracting from the arcane literature of economics basic concepts, awareness of controversies over their application, insights into puzzles and patterns, and an intellectual thrust to jump between economics and strategy.

ECONOMIC CRISES AND STRATEGIC COMPETITION

While assuming a high probability of strategic competition, it is important to consider what might dampen the intensity of competition. Warning of a "Thucydides trap" in the rivalry between rising and ruling powers, Graham Allison emphasizes that most long-term competition ends in conflict and that resolution short of war requires "huge adjustments in attitudes and actions" of those involved.[3] One important cause of adjustments may lie in the realm of economics. Short-term financial crises and the long-term spillover effects represent internal material circumstances most likely to bring about a period of either Chinese or American relative quiescence in their external strategic behavior.

Financial crises may take a multitude of forms, including runs on banks or currencies, burst bubbles in asset markets, and unsustainable rises in external debt or public deficits. Historically, financial crises have had a dramatic impact on the domestic stability and strategic solvency of great power regimes, as in the Soviet Union at the end of the Cold War. Nevertheless, for decades before the shock of 2007–2008, financial crises had migrated to the margins of economic theory. The pre-2008 consensus among theorists was that financial markets in advanced economies no longer were prone to catastrophic crashes. Both Charles Kindleberger and Hyman Minsky, however, provided deeper insight into the instability of financial markets, pointing out that economic success holds the seeds of its own demise.[4]

Such a perspective clarifies the experience of East Asia in recent decades. The astonishing ascent of the Japanese economy from the 1950s to the 1980s gave way to the financial crisis and dismal economic effects that have persisted for more than two decades. The rise of South Korea from poverty in the 1950s to prosperity fifty years later was punctuated by several financial crises. The financial crisis of 1997–1998, which began in Thailand and spread to Indonesia, Malaysia, the Philippines, and South Korea, was preceded by inflows of capital from foreign investors who thought they saw evidence of new miracles in the making in emerging economies. When the inflows became outflows, booms turned to busts.

After the shock of 2008 economists scrambled to build a framework to understand financial crises and predict the next one. This framework stresses credit booms as the best indicators of looming financial crises. If and when a credit bust happens, it might trigger an economic recession or reinforce other phenomena that drive downturns. When financial crises become intertwined with economic recessions, downturns tend to be unusually deep and long. As banks and other institutions are bailed out, as producers and consumers deleverage their balance sheets, and as governments resort to fiscal and monetary counteraction, the burden of debt passes to the public sector. Beyond a certain level, the higher that debt reaches in relation to the gross domestic product (GDP), the lower the postcrisis growth. The controversial parts in this story are the short-term effectiveness of fiscal and monetary stimulus in curtailing downturns and the long-term relationship between debt levels and growth rates after recovery from the recession.[5]

Appraisals of strategic competition in the Asia-Pacific region should relate this framework to Japan, the United States, and China. In the Japanese case, when the bubbles in equity and real estate markets burst in the late 1980s, that created paralysis in the banking world and damaged highly leveraged corporate balance sheets during the 1990s. Real annual GDP growth, which had averaged about 4 percent between the mid-1970s and the late 1980s, decreased to an average of less than 1 percent in the two decades after 1990. What is more, public debt rose to a higher level in relation to GDP than in any other advanced economy in the world.[6]

Two decades of dismal growth left Japan with a badly deteriorated competitive posture compared to China. In 1990 Japanese military expenditure amounted to 2.4 times that of China. But since Japan allocated only 1 percent of a slowly growing national income to defense and China boosted military outlays to more than 2 percent of a national income whose growth rate was often ten times greater, Chinese military expenditure reached 2.7 times that of Japan in 2012.[7] This disparity portends greater Japanese reliance on the United States to counterbalance China.

The United States, however, has major economic and fiscal problems of its own. After the second half of the 1990s, an accelerated expansion of debt occurred in American financial institutions and households, most prominently in real estate mortgages. The credit boom on the flip side of the debt explosion resulted not only from exuberant risk taking by loosely regulated banks and unregulated shadow banks but also from capital flows from China and other countries in Asia whose manipulation of exchange rates and repression of rates of return on savings translated into an increased current-account deficit of the United States.[8] The housing bust in 2007 and the shadow-banking implosion in 2008 led to the worst economic decline since the Great Depression. Six years after this crisis began, US national income was only 6 percent greater than its precrisis level in real (inflation-adjusted) terms.

As the American private sector deleveraged, government debt skyrocketed. Over three fiscal years the deficit ran between 8 and 10 percent of GDP.[9] The Budget Control Act of 2011 slammed on the brakes, and the military budget absorbed much of the fiscal crunch as sequestration left Medicare, Medicaid, and Social Security untouched. In the absence of faster economic growth or the political will to raise taxes and cut entitlements, increases in spending make an even more intense fiscal crisis highly likely for the United States in the 2020s and 2030s.

In such a crunch, the greatest trade-off of strategic consequence will involve military and medical spending. The odds are not favorable for the defense budget. Consider the long-term pattern. In 1960, during the intense competition with the Soviet Union and before Medicare, the United States spent about 10 percent of GDP on defense, but federal spending on health care was less than 0.3 percent of GDP. By fiscal year 2013 budgetary outlays on health care had caught up to military spending.[10] Much economic analysis has been done on why medical spending has spiraled upward.[11] Political factors are important as well. American politicians often have been willing to cut military spending; they have not been willing and able to reverse the momentum of government medical spending at any time since the advent of Medicare.

Demographic trends will exacerbate the looming fiscal crunch. At the end of FY13 the Congressional Budget Office projected that in twenty-five years, with an aging population, federal spending on health care, Social Security, and interest on the national debt will rise to 19 percent of GDP.[12] If revenues follow the historical pattern of rarely going above 20 percent of GDP, outlays to support competition with China will be squeezed. The likelihood of US inactivity in the face of Chinese assertiveness would be high. It is unsurprising that Chinese strategic thinkers highlight the opportunities to take advantage of American fiscal and financial difficulties.[13]

The Chinese are themselves not immune to such difficulties. To be sure, their policy makers were able to maneuver buoyantly through the Asian financial crisis of 1997–1998 and quasi-global recession a decade later. But the stimulus measures introduced by the Chinese government in late 2008 to ward off the recession greatly increased the chances of a financial crisis in China several years later. Rather than step up central government expenditure and drive its budget into deficit, the leadership in Beijing directed state-owned banks and local governments to finance a spending spree on infrastructure and industrial expansion. The resulting increase of credit in relation to GDP was extraordinary, beyond even what had taken place in Japan from 1985 to 1990, in South Korea from 1994 to 1998, and the United States from 2002 to 2007.[14] By 2013, there was much discussion among financial analysts and journalists about a possible hard landing of the Chinese economy.[15] And International Monetary Fund representatives in Beijing pointed out that if one added local government fiscal activities to central government budgetary accounts, the augmented fiscal deficit of China was about 10 percent of GDP in 2012. There was much less fiscal space to soften a landing than there had been in 2008.[16]

If a hard landing occurs and is followed (as Japan and the United States experienced) by a prolonged shortfall in economic growth, it could threaten the grip of the Communist Party. But gauging possible international strategic consequences of such events is not straightforward. On one hand, given the conventional wisdom that the legitimacy of the communist regime rests on two pillars, economic performance and nationalism, one might anticipate that damage to the first pillar would lead the regime to reinforce the second pillar and act more aggressively in the international arena. On the other hand, given that such challenges to regime survival concentrate the minds of communist leaders, détente may be more probable in strategic competition.

Assessments must be attuned to paradoxes with dialectical twists or yin-yang dynamics. For example, though the interdependence of economies may help prevent war, high rates of economic growth may increase intense strategic competition and diminished growth dampen it. Furthermore, entwined financial crises and economic slumps may emerge from times of financial stability and economic success. But any periods of crisis and stagnation may give rise to major economic reforms. The regeneration of economic growth, if it sprouts from the seeds of reform, is the best solution to financial crises and fiscal crunches.

It is not easy to bring the seeds of reform to fruition. Amid the swirl of general theoretical ideas advanced by economists, policy makers must judge which apply in specific circumstances. Political leaders must have the will and skill to overcome vested interests opposed to reforms that in Asia usually involve removing impediments to market forces. And, not least, the payoffs of whatever reforms are implemented rely on both the potential of the economy and the extent to which reforms tap into key elements of that potential.

ECONOMIC POTENTIAL AND STRATEGIC COMPETITION

The connection between economic prowess and long-term strategic success was highlighted by Paul Kennedy in *The Rise and Fall of the Great Powers*.[17] As a historian, Kennedy did not give a theoretical explanation of economic growth. He could have benefited from a growth-accounting framework laid out decades earlier by Robert Solow. To analyze the growth of national output, Solow differentiated the contributions of capital investment and technical change. Applying his theory to data on the US economy from 1909 to 1949, he concluded that technical change accounted for the bulk of nonagricultural growth.[18]

An academic growth industry was born and economists sought to measure inputs at various times and places: capital inputs, qualitative improvement in capital equipment, labor inputs, and qualitative improvements in the workforce (human capital). When those growth inputs were added up, they fell short of output to either a greater or a lesser degree. The residual was termed *total factor productivity* (TFP), which under analysis turned out to represent more than technological change. Organizational change, new business models, and other means of achieving efficiency were valuable. In cases of transition from an economy dominated by agriculture, structural change added to TFP as workers moved from farms to more productive factory jobs.

Controversies exist over assumptions, measurements, categories, and relationships among variables. Dale Jorgenson and his disciples challenged Solow and his followers on the grounds that capital inputs, properly measured, make a greater contribution than total factor productivity to growth.[19] That controversy may be important to those engaged in the strategic assessment of tradeoffs between military power and economic growth. Capital consumed in the production of military power is capital that cannot be invested in expanding economic capacity. As the Soviet experience revealed by the mid-1970s, if additional output has to be generated by the accumulation of capital, increased military expenditure decreases economic growth.[20] But to the extent that high rates of growth can be generated by innovations that do not involve high capital costs, the trade-off between military power and economic growth is attenuated. An economy that benefits from TFP growth of 2 percent of GDP can accommodate increased military expenditure in a way that an economy that gets a more modest TFP contribution to output cannot, without a sacrifice in long-term economic growth.

Exercises in growth accounting require reliable data. Where deficiencies in data exist, such as in the People's Republic of China (PRC), there is abundant controversy. Economists most skeptical of Chinese data—notably Alwyn Young and Harry Wu—recalculate the annual TFP growth rate at less than 1.5 percent for most periods except after China joined the World Trade Organization (WTO) in 2001.[21] That is similar to the TFP growth rate of the United States from 1995 to 2007 and less than expected for a developing country as it opened to foreign trade and assimilated advanced technology from abroad. By contrast, growth accounting done with less reworking of PRC national-level data has found TFP growth rates of more than 3.5 percent per year in China since the late 1970s. Such extraordinary rates outpaced those anywhere else in Asia. A 2012 study by Loren Brandt and coauthors, using more reliable firm-level data, has strongly reinforced the case for high rates of Chinese TFP growth.[22]

Growth accounting provides a conceptual framework for assessing economic potential and analyzing policies to achieve it. Chinese economic policies align more closely to the framework than do those of the competition. Even though the WTO surge of 2001–2008 was exhilarating, Chinese policy makers realized that capital inputs were generating diminishing marginal returns for state-owned enterprises (SOEs) and labor inputs would diminish as the population aged. To maintain high GDP growth rates, TFP growth would have to make an enhanced contribution.

One key to TFP growth is technology. In seeking an indigenous capability to innovate at the technological frontier, the Chinese have expanded and reoriented their educational and research-and-development systems. While pursuing that aim, China has been ensuring that its economy absorbs technology from advanced nations.[23] Much more than Japan and India have, China has long welcomed foreign direct investment, but with a twist, ensnaring foreign companies in joint ventures in which their price of access is transfer of their technology. As foreign companies have been twisted harder, Beijing has been running the risk of scaring them away. Hedging against that risk, China has sought the acquisition of companies abroad to gain technological know-how. Another way to get such knowledge is by inducing Chinese entrepreneurs and researchers to return home from overseas. Less conventional but more important is cyber-theft of foreign technology. In a speech delivered in 2011, the head of US Cyber Command stated, "What's been going over the last few years in the networks is the greatest theft that we've seen in history. What we're losing in intellectual property is astounding."[24]

A second key to future TFP growth in China is the further movement of labor from low-productivity work in the countryside to high-productivity jobs in cities. By 2013 economists were predicting that the Chinese were quickly approaching the Lewis turning point, when surplus rural labor would no longer be available and wages would increase. But Li Keqiang, the first premier to hold a doctorate in economics, highlighted the scope for greater urbanization. Whereas the urban population in advanced economies is typically three-quarters of the total, the fraction in China is only half.[25]

Along with the movement of workers to factories, a third key to TFP growth has been the long-term shift of production from relatively inefficient SOEs to more efficient private-sector or mixed-ownership businesses. At the Third Plenum in 2013, new president and Communist Party secretary Xi Jinping stated his interest in accelerating this shift. The Xi-Li team was aware that private enterprise, despite having long been disadvantaged by state policies, produces higher rates of return on capital, more employment, and greater TFP levels than SOE. Allowing the private sector to both compete on an equal footing with state businesses and enter SOE-dominated sectors could give a major boost to TFP growth rates.

In addition to a long-term focus on a growth-accounting framework, Chinese policy makers keep an eye on a national-income-accounting framework that includes investment, consumption, government purchases of goods and services, and exports less imports. Under this framework, as the Chinese economy went through a WTO surge from 2001 to 2008 and the post-2008 stimulus, investment expanded to more than 50 percent of GDP, the current-account surplus expanded to almost 10 percent, and household consumption fell to 35 percent—half that of the United States and two-thirds that of Japan and India. Even before the quasi-global economic crisis of 2008, the premier at the time, Wen Jiabao, announced that the Chinese economy had become "unstable, unbalanced, uncoordinated, and unsustainable."[26] The disparity between investments and consumption rested on links among large banks, SOEs, and local governments. Although government policies had shaped this iron triangle, once in place it was hard to change.

Thus the new Xi-Li leadership sought simultaneously to enhance long-term productivity growth, engage in medium-term rebalancing, and avert a short-term financial crisis. Its reform program, publicly unveiled after the Third Plenum of November 2013, included proposals that would raise TFP by moving capital and labor to more productive uses and would boost consumption as a proportion of national income. The most notable ways to achieve these ends are liberalizing interest rates, especially on deposits in banks; expanding equity and bond markets; eliminating price controls on factor inputs that in effect subsidize investment by SOEs; opening SOE-run sectors to greater competition; granting secure property rights to farmers; relaxing the notorious *hukou* household-registration system, which controls movement from rural to urban areas; and funding extensive social-welfare programs.[27] Such measures pose challenges to the iron triangle. Absent wide spreads between deposit rates and lending rates set by policy makers, large banks would have to improve their capability to assess and manage risk. Without profits guaranteed by government subsidies, SOEs would have to become much more efficient. And local governments, without the ability to fund themselves by expropriating land from farmers and circumventing regulations on borrowing, must hope that the central government offers them new sources of revenue and shoulders some of their responsibility for social-welfare spending.

The success of these ambitious reforms is not assured. On one hand, if fully implemented by Xi Jinping's target date of 2020, they may precipitate rather than avert a financial crisis, especially if domestic liberalization is followed, in a quest to make the renminbi a leading international currency by an effort to open up Chinese capital markets to external financial flows.[28] On the other hand, this program may be only fitfully implemented. Embedded in the iron triangle are vested interests in existing ways of doing business—ways that have produced extraordinary success as well as enriched and empowered Communist leaders while enabling them to maintain an ideological fig leaf of socialism. Many elements of the Xi-Li program have appeared on the central government agenda for years, but Hu Jintao and Wen Jiabao apparently were unable to overcome opposition of vested interests. During his first year in office, Xi already has demonstrated the skill in political maneuver and will for economic reform that previous Chinese leaders have lacked. He is motivated by an awareness that successful economic reforms pay dividends in soft power (the image of China around the world), sticky power (the attraction of an expanded Chinese consumer market to foreign companies with political influence in their home countries), and hard power (increases in military capabilities).[29]

The implications of a new wave of economic reforms for long-term strategic competition are sobering. Chinese policy makers seek to maintain an annual GDP growth rate above 7 percent, the rate calculated to keep urban unemployment at 4 percent or less and thus minimize chances of domestic instability. An aging Chinese population will make that growth rate hard to sustain by 2030. Therefore, assume that the successful execution of the Xi-Li reforms in the coming decade makes possible real annual growth of 6 percent on average for a generation. Moreover, suppose that the United States returns to its previous growth rate of 3 percent.[30] In twenty-five years, if one uses 2012 purchasing-power parity as a basis of comparison, Chinese GDP might be more than 50 percent that of the United States. In 1953, a secret analysis by the Eisenhower administration assumed that the Soviet economy would grow at an annual rate of 6 percent and the US economy would increase by 3.6 percent. It projected that because the Soviet Union started with a small fraction of the US economy (a much smaller fraction than China did sixty years later), the aggregate gap between the Soviet and US economies would not close, despite the assumed difference in growth rates.[31] Within the economic context of strategic competition, China presents a potential challenge to the United States that the Soviet Union was never capable of mounting.

Japanese prospects are even more sobering, not least because of a dismal demographic outlook. Because of the odd combination of low fertility and low female participation in the labor force, the working-age population of Japan peaked around 1990 and will decline far into the twenty-first century. The Chinese working-age population peaked two decades after Japan's, and its decline will be less steep. While the Chinese working-age population by 2050 will fall to its 2000 level, the Japanese labor force will be 40 percent less than in 2000.[32] A rise in quality can make up for a drop in numbers, but jumps in educational attainment in Japan are phenomena of the past.

Moving from labor inputs to capital inputs in a growth-accounting framework, one also does not find much compensation for demographic decline. On the contrary, an aging society means the amount of Japanese household saving that can finance investment will continue to fall, and an additional 7 percent of GDP might be diverted to pay for health care and pensions by 2030.[33] Despite high levels of corporate savings, business firms have preferred padding balance sheets or investing overseas to adding capital stock at home. Private-sector capital formation as a percentage of GDP fell by roughly a third in the decades after 1990, along with diminishing returns on capital invested.[34]

Except for an upswing when Koizumi Junichiro was the prime minister, from 2001 to 2006, growth rates of TFP also stagnated.[35] While the Japanese spend an impressive amount on research and development, some of their best-performing companies of the past have become less innovative and have lost their competitive edge in overseas markets. Entrepreneurship is not much in evidence, perhaps because it is harder to start a business in Japan than in other advanced economies.[36]

The reallocation of resources from unproductive to productive firms and sectors does not go smoothly either, as one can see from Japanese economic policy in the recent dismal decades. Whereas both China and India undertook vigorous reform efforts in the 1990s in response to crises, Japan reacted to its crisis sluggishly. Banks kept rolling over nonperforming loans to zombie firms, thereby impeding the productivity-enhancing exit and entry of businesses in competitive markets.[37] Not until Takenaka Heizo became finance minister under Koizumi did the government impose reform on the banks and restructuring on their insolvent corporate customers. Koizumi then tried to carry reform further, with a plethora of initiatives, too few of which focused on increasing TFP growth.[38] His successor, Abe Shinzo, proved to be "rather clueless."[39]

When Abe became the prime minister again in December 2012, he was a different leader, not just because he had learned from failure but also because he had a much better grasp of the connection between economics and strategy. His main objective remained unchanged: restoring Japan as a great power capable of projecting military force for collective defense as well as self-defense. The urgency of that objective increased with China's assertiveness since 2008, after Abe had served his first stint as premier, and the emergence of the Senkaku Islands as a major point of conflict before he took office for a second time. The security posture envisioned by Abe required amending the constitution, a change that the majority of Japanese people have opposed. For Abe, in the short term, economic reforms that reignite growth might well give him political capital to parley into more support for his strategic design. In the long term, sustained economic resurgence would markedly strengthen Japan in its strategic competition with China.[40]

The Abe economic-reform initiative aimed three arrows at a target of real GDP growth of 2 percent annually over ten years. The first two arrows—fiscal stimulus and monetary easing—make sense in a national-income-accounting framework. Fiscal stimulus has the first-order effect of growing the government portion of GDP and then multiplier effects on consumption and investment. Monetary easing, on a scale perhaps three times that of quantitative easing (relative to GDP) by the US Federal Reserve, stimulates consumption by creating inflationary expectations, increases investment by making more cheap credit available, and boosts exports over imports by depreciating the yen. There are limits, however, to how far Abe can safely shoot these first two arrows, owing to concerns of investors about the high level of government debt and the "step in the dark" that unconventional monetary policy represents.[41] Japan can ill afford the flight of foreign investors from Japanese financial markets.

Economists generally agree that short-term fiscal and monetary stimuli do not raise long-term growth rates. For that to happen, a third arrow must work through the growth-accounting framework. Prime Minister Abe hopes to increase labor inputs with greater participation in the workforce by women and retirees. Permitting more immigrants to enter the labor market could result in valuable contributions to growth, but Abe shows little interest in challenging the Japanese cultural aversion to an influx of foreigners. To increase capital inputs, he is setting up special economic zones free of the regulations that pervade the Japanese economy. But he is shying away from tax policies that would strong-arm, rather than jawbone, corporations to turn their accumulated savings into an augmented capital stock.[42] To regenerate TFP growth, the reform program will have to strike at the misallocation of factor inputs holding back the Japanese economy. That would involve relaxing labor-market restrictions and removing tariff protection of the inefficient agricultural sector. Such measures, which expose vulnerable groups to free-market forces, are controversial in Japanese politics.

Even if Abe has the political will and skill to circumvent obstacles to restructuring the Japanese economy, a big constraint remains: the long-term economic potential of Japan is far less than that of China. Suppose Abe meets his goal of 2 percent growth, and suppose China grows at a rate of 6 percent for the next generation. The size of the Chinese economy would be five to seven times that of the Japanese economy by the late 2030s. If the United States manages to grow at a 3 percent rate over the same period, the Chinese economy would be larger (in 2012 purchasing-power-parity terms) than the American and Japanese economies combined.

Unlike Japan, India has the economic potential to compete with China in the long run. Of all the major economies of the world, only the Indian economy can reap a major demographic dividend from a rising working-age population in the coming decades. India can draw on a high rate of household savings, though chronic government budget deficits divert some savings from productive investment. A remarkable feature of the Indian economy has been the development of a service sector with TFP growth not typically seen in a nation at a low level of development.[43] There remains ample scope for structural change. The agricultural sector employs roughly 60 percent of the labor force, and its productivity has been stagnant since the end of its Green Revolution, when new crop varieties and farming techniques increased production. If productivity in the industrial sector could reach the levels enjoyed by the service sector and if labor could be reallocated more quickly from agriculture, the GDP growth rate in India could exceed that of China in the future.

The greatest impediment to improved performance is the political system. Though praiseworthy for sustaining democracy in a large, heterogeneous society, India has fallen short of China in improving the economic condition of the bulk of the population. Since the Manmohan Singh reforms of the 1990s, the commitment to economic liberalization has slackened, especially under Singh's lackluster leadership as prime minister when the Indian economy fell hard from peak growth after 2010. India is still less open to foreign trade and direct investment than China. It maintains protective labor legislation that has prevented industry from making faster economic gains. Moreover, the Indian central government has been deficient in basic tasks such as educating the masses and building infrastructure.[44]

Three developments could bring about a renewal of economic reform in India's political system. One would be a financial crisis, as in 1991. The second would be the ascendancy to national leadership of chief ministers of states that have been successful with economic reform. Narendra Modi, chief minister of Gujarat and standard-bearer of the Bharatiya Janata Party, is the most charismatic and controversial of those potential national leaders.[45] The third would be the political stimulus of a much-intensified strategic competition with China.

The most challenging issue that a strategic assessment of the Asia-Pacific region in the coming decades must address is the prospects for the American economy. I approach that issue from a different vantage point than I adopt for the other major powers of the region. China, Japan, and India are in a catch-up mode to a United States that has long been perched at the production-possibility frontier. Economists have proposed various hypotheses about the dynamics of the catching-up process. One is the convergence hypothesis, which in its simplest form suggests that the farther a developing nation is from the frontier, the faster it should be able to advance toward it, once certain conditions for economic growth are in place. Another hypothesis advanced by Barry Eichengreen and his colleagues is that the journey to the frontier is likely to stall in a middle-income trap when nations attain a per capita income of $15,000 to $17,000 (in 2005 US dollars on a purchasing-power-parity basis).[46] But what if a strategic competitor such as China does not stall and approaches the production-possibility frontier at high speed? Can the United States change the terms of the game?

THE NEW FRONTIER

The economic key to becoming and remaining a leading global power since the nineteenth century has been securing and maintaining a position on the production-possibility frontier. That, in turn, has required being on the leading edge of technological innovation in those areas in which the potential TFP growth was greatest. When neoclassical economists conduct quantitative analyses of growth on the frontier, they depict impersonal market forces operating without much drama. But a closer examination reveals a turbulent history of entrepreneurs pushing the frontier and generating what Joseph Schumpeter described as the creative destruction of capitalism.[47]

A chronology of great economic leaps would list clusters of innovation at certain times in certain places. Inventors put together new technologies, and entrepreneurs combine the technologies with new business models. The pushing out of the production-possibility frontier means increases in TFP and output growth. In this process, one can find patterns in the clusters of innovation: two industrial revolutions and one information revolution over the last two centuries of modern economic growth.

Those who want to connect economics and strategy will see that the foremost global power of the nineteenth century, Great Britain, was on the leading edge of the first industrial revolution, while other powers sought to catch up. As the second industrial revolution developed from the late nineteenth to mid-twentieth centuries, Britain lost out to ambitious competitors, above all to Americans. The leading-edge dynamics of the second industrial revolution constituted the material dimension of the American rise to global power by the mid-twentieth century. Other nations, European and then East Asian, played the catch-up game fairly well in the second half of the century but not as well as China has played it recently. Just as Paul Kennedy pointed to signs of a relative decline in America in 1987, the information technology (IT) revolution gave the United States the opportunities to stake out a leading position on a new production-possibility frontier.

There are puzzles in the latest iteration of the revolutionary pattern. Is the IT revolution as powerful as the second industrial revolution?[48] Has it already played out, or is more innovation to come? Strategic assessments should examine these puzzles and patterns. The consumers of assessments must be reminded that successful competition with China requires the United States to focus on maximizing TFP. Pushing to a new technological frontier as fast as possible will force China to start playing the catch-up game all over again.

Developing an intellectual basis for a long-term assessment that connects economics and strategy requires awareness of the parallels between economic and strategic challenges. During several decades the US military has been on the technological edge of precision-strike weapons and China has been catching up faster than thought.[49] Like the *production-possibility* frontier, the United States has a strong incentive to pursue the *strategic-possibility* frontier. The growth-accounting framework of economic theory can be adapted to generate strategic benefits from military inputs. From this perspective, big leaps to a new strategic frontier are likely to occur less from greater capital investment in weapons platforms and further labor inputs of trained military personnel than from technological, organizational, and conceptual innovation.[50]

The concept of the game changer may be a cliché, but it has significance for the theory and practice of economics and strategy. Game changing must first be grasped, which involves theory, and then practiced. In strategy and economics, intellectual assessment represents only a point of departure. Creative entrepreneurs must blaze the trail to new frontiers.

NOTES

1. Robert B. Strassler, ed., *Thucydides: A Comprehensive Guide to the Peloponnesian War* (New York: Simon and Schuster, 1996), 43.

2. Evan A. Feigenbaum and Robert A. Manning, "A Tale of Two Asias: In the Battle for Asia's Soul, Which Side Will Win—Security or Economics?" *Foreign Policy*, October 31, 2012, http://www.foreignpolicy.com/articles/2012/10/30/a_tale_of_two_asias.

3. Graham Allison, "Thucydides's Trap Has Been Sprung in the Pacific," *Financial Times*, August 21, 2012, http://www.ft.com/intl/cms/s/0/5d695b5a-ead3-11e1-984b-00144feab49a.html#axzz30IVoY9iZ.

4. Charles P. Kindleberger, *Manias, Panics, and Crashes: A History of Financial Crises* (New York: Basic Books, 1978); Hyman Minsky, *Can "It" Happen Again? Essays on Instability* (Armonk, NY: M. E. Sharpe, 1982).

5. Carmen M. Reinhart and Kenneth S. Rogoff, *This Time Is Different: Eight Centuries of Financial Folly* (Princeton, NJ: Princeton University Press, 2009); Reinhart and Rogoff, "From Financial Crash to Debt Crisis," *American Economic Review* 101 (August 2011): 1676–1706; Carmen M. Reinhart, Vincent R. Reinhart, and Kenneth S. Rogoff, "Public Debt Overhangs: Advanced-Economy Episodes Since 1800," *Journal of Economic Perspectives* 26 (Summer 2012): 69–86; Oscar Jorda, Moritz Schularick, and Alan M. Taylor, "Financial Crises, Credit Booms, and External Imbalances: 140 Years of Lessons," *IMF Economic Review* 59 (2011): 340–78; Alan M. Taylor, "The Great Leveraging," National Bureau of Economic Research Working Paper 18290, August 2012.

6. For the demand side since the 1980s, see Richard C. Koo, *The Holy Grail of Macroeconomics: Lessons from Japan's Great Recession* (Singapore: John Wiley and Sons, 2009). On the supply side, see Takeo Hoshi and Anil Kashyap, "Why Did Japan Stop Growing?" National Institute for Research Advancement, January 2011.

7. Military spending figures are from the Stockholm International Peace Research Institute's Military Expenditure Database, 1988–2013, http://www.sipri.org/research/armaments/milex/milex_database/milexdata1988-2012v2.xsls/view, and GDP statistics are from the World Bank, "GDP Growth (Annual %)," http://data.worldbank.org/indicator/NY.GDP.MKTP.KD.ZG.

8. David Luttrell, Harvey Rosenblum, and Jackson Thies, "Understanding the Risks Inherent in Shadow Banking: A Primer and Practical Lessons Learned," *Staff Papers*, no. 18 (November 2012); C. Fred Bergsten and Joseph E. Gagnon, "Currency Manipulation, the US Economy, and the Global Economic Order," Peterson Institute for International Economics Policy, Policy Brief 12-25, December 2012; Nicholas R. Lardy, "Financial Repression in China," Peterson Institute for International Economics Policy, Policy Brief 08-8, September 2008.

9. See Timothy Taylor, "Goldilocks Fiscal Policy: Just About Right," *Conversable Economist* blog, November 13, 2013, http://conversableeconomist.blogspot.com/2013/11/goldilocks-fiscal-policy-just-about.html.

10. Calculations are from data on the US Office of Management and Budget website at http://www.whitehouse.gov/omb/budget/historicals; the website US Government Spending, at http://www.usgovernmentspending.com/year_download_1960USbn_15bs2n#usgs302 and http://www.usgovernmentspending.com/year_download_2013USbn_15bs2n#usgs302; and Congressional Budget Office, "Monthly Budget Review—Summary for Fiscal Year 2013," November 7, 2013.

11. Kenneth J. Arrow, "Uncertainty and the Welfare Economics of Medical Care," *American Economic Review* 53 (December 1963): 141–49; William J. Baumol, *The Cost Disease: Why Computers Get Cheaper and Health Care Doesn't* (New Haven, CT: Yale University Press, 2012); Amy Finkelstein, "The Aggregate Effects of Health Insurance: Evidence from the Introduction of Medicare," *Quarterly Journal of Economics* 122 (February 2007): 1–37; Alan M. Garber and Jonathan Skinner, "Is American Health Care Uniquely Inefficient?" *Journal of Economic Perspectives* 22 (Fall 2008): 27–50.

12. Congressional Budget Office, "The 2013 Long-Term Budget Outlook," September 17, 2013.

13. Huang Renwei, "China Can Take Advantage of the Limitations of the US," *Xinhua* [China View], December 20, 2012, http://news.xinhuanet.com/world/2012-12/20/c_124122632.htm [in Chinese].

14. See graph in Edward Chancellor and Mike Monnelly, "Feeding the Dragon: Why China's Credit System Looks Vulnerable," GMO white paper, January 2013, p. 3.

15. "China: Rising Risks of Financial Crisis," Nomura Global Economics Asia special report, March 15, 2013; Andy Xie, "No Sugarcoating It: A Hard Landing Is Likely," *Caixin*, November 21, 2013.

16. International Monetary Fund, "People's Republic of China: 2013 Article IV Consultation," July 2013, p. 13.

17. Paul M. Kennedy, *The Rise and Fall of the Great Powers: Economic Change and Military Conflict from 1500 to 2000* (New York: Random House, 1987).

18. Robert M. Solow, "Technical Change and the Aggregate Production Function," *Review of Economics and Statistics* 39 (August 1957): 312–20.

19. D. W. Jorgenson and Z. Griliches, "The Explanation of Productivity Change," *Review of Economic Studies* 34 (July 1967): 249–83; Dale W. Jorgenson, ed., *The Economics of Productivity* (Cheltenham, UK: Edward Elgar, 2009).

20. Myron Rush, "Guns over Growth in Soviet Policy," *International Security* 7 (Winter 1982–1983): 167–79.

21. Alwyn Young, "Gold into Base Medals: Productivity Growth in the People's Republic of China During the Reform Period," *Journal of Political Economy* 111 (December 2003): 1220–61; Harry X. Wu, "Accounting for China's Growth in 1952–2008: China's Growth Performance Debate Revisited with a Newly Constructed Data Set," Research Institute of Economy, Trade and Industry, January 2011; "China's Growth

and Productivity Performance Debate Revisited: Accounting for China's Sources of Growth in 1949–2010 with a Newly Constructed Data Set" (paper presented at the University of Western Australia Chinese Economy Workshop, April 3–4, 2013).

22. Barry Bosworth and Susan M. Collins, "Accounting for Growth: Comparing China and India," *Journal of Economic Perspectives* 22 (Winter 2008): 45–66; Dwight H. Perkins and Thomas G. Rawski, "Forecasting China's Economic Growth to 2025," in *China's Great Economic Transformation*, ed. Loren Brandt and Thomas G. Rawski (New York: Cambridge University Press, 2008), 839; Jungsoo Park, "Projection of Long-Term Total Factor Productivity Growth for 12 Asian Economies," Asian Development Bank, October 2010, pp. 7–9; Loren Brandt, Johannes Van Biesebroeck, and Yifan Zhang, "Creative Accounting or Creative Destruction? Firm-Level Productivity Growth in Chinese Manufacturing," *Journal of Development* Economics 97 (2012): 339–51.

23. Kirsten Bound, Tom Saunders, James Wilsdon, and Jonathan Adams, *China's Absorptive State: Research, Innovation and the Prospects for China-UK Collaboration* (London: Nesta, 2013).

24. Keith B. Alexander, quoted in Zachary Fryer-Biggs, "Passive Defenses Not Enough: U.S. Cyber Commander," *Defense News*, September 14, 2011, http://www.defensenews.com/print/article/20110914/DEFSECT04/109140315/Passive-Defenses-Not-Enough-U-S-Cyber-Commander.

25. W. Arthur Lewis, "Economic Development with Unlimited Supplies of Labor," *Manchester School* 22 (May 1954): 139–91; Mitali Das and Papa N'Diaye, "Chronicle of a Decline Foretold: Has China Reached the Lewis Turning Point?" International Monetary Fund, January 2013; Li Keqiang, "Promoting Coordinated Urbanization—an Important Strategic Choice for Achieving Modernization," September 9, 2012, http://www.primeeconomics.org/wp-content/uploads/2013/06/Li-Keqiang-China-urbanization-speech.pdf.

26. Quoted in Ling Zhu, "Premier: China Confident in Maintaining Economic Growth," *Xinhua* [China View], March 16, 2007, http://news.xinhuanet.com/english/2007-03/16/content_5856569.htm.

27. "The Decision on Major Issues Concerning Comprehensively Deepening Reforms," *China Daily*, November 18, 2013.

28. On the dangers of international hot money flows, see Hélène Rey, "Dilemma Not Trilemma: The Global Financial Cycle and Monetary Policy Independence" (paper presented at Federal Reserve Bank of Kansas, Economic Policy Symposium, Jackson Hole, Wyoming, August 2013); Victor Shih and Susan Shirk, "To Renminbi or Not to Renminbi: Why China's Currency Isn't Taking Over the World," *Foreign Policy*, October 18, 2012, http://www.foreignpolicy.com/articles/2012/10/18/to_renminbi_or_not_to_renminbi.

29. For more on Xi and soft power, see his justification of reform reported in Lu Hui, "Xi Explains China's Reform Plan," *Xinhua*, November 15, 2013, http://news.xinhuanet.com/english/china/2013-11/15/c_132891949.htm. On sticky power, see Walter Russell Mead, *Power, Terror, Peace, and War: America's Grand Strategy in a World at Risk* (New York: Random House, 2004), chap. 2.

30. Malhar Nabar and Papa N'Diaye, "Enhancing China's Medium-Term Growth Prospects: The Path to a High-Income Economy," International Monetary Fund, October 2013; Robert J. Gordon, "Is US Economic Growth Over? Faltering Innovation Confronts the Six Headwinds," National Bureau of Economic Research, August 2012.

31. Bradford A. Lee, "American Grand Strategy and the Unfolding of the Cold War, 1945–1961," in *Successful Strategies: Triumphing in War and Peace from Antiquity to the Present*, ed. Williamson Murray and Richard Sinnreich (New York: Cambridge University Press, 2014), 399.

32. See the graph in David Wessel, "The Demographics Driving Nations' Wealth," *Wall Street Journal*, August 12, 2010.

33. Richard Jackson, "Japan's Demographic End Game," Center for Strategic and International Studies, November 21, 2013.

34. International Monetary Fund, "Japan Sustainability Report," 2011, pp. 5–6; Kyoji Fukao and Hyeog Ug Kwon, "Growth Strategy Task Force White Paper," American Chamber of Commerce in Japan, June 2011.

35. One estimate puts the average of TFP growth in Japan for 1990 to 2010 as low as 0.1 percent. Asia Productivity Organization, *APO Data Book 2012* (Tokyo: Keio University Press, 2012), 76.

36. Ingo Beyer von Morgenstern, Peter Kenevan, and Ulrich Naeher, "Rebooting Japan's High-Tech Sector," *McKinsey Quarterly*, June 2011, http://www.mckinsey.com/insights/high_tech_telecoms_internet/rebooting_japans_high-tech_sector; Anthony Fensom, "Japan's Lost Art of Innovation," *Diplomat*, October 10, 2012, http://www.thediplomat.com/2012/10/japans-lost-art-of-innovation.

37. Richard J. Caballero, Takeo Hoshi, and Anil K. Kashyap, "Zombie Lending and Depressed Restructuring in Japan," *American Economic Review* 98 (2008): 1943–77.

38. Takeo Hoshi and Anil K. Kashyap, "Will the US and Europe Avoid a Lost Decade? Lessons from Japan's Post Crisis Experience" (paper presented at Fourteenth Jacques Polak Annual Research Conference, November 2013), pp. 8–9, 41.

39. "Once More with Feeling: Japan and Abenomics," *Economist*, May 18, 2013, p. 24.

40. Ibid. See also Gerard Baker and George Nishiyama, "Abe Sees a Resurgent Japan Leading Response to China," *Wall Street Journal*, October 25, 2013; and Linda Sieg, "Analysis: Japan PM to Push Security Agenda Next Year with Fresh Urgency," Reuters, December 14, 2013.

41. Raghuram Ragan, "A Step in the Dark: Unconventional Monetary Policy After the Crisis" (Andrew Crockett Memorial Lecture, Bank for International Settlements, June 23, 2013).

42. Martin Wolf, "Japan's Unfinished Policy Revolution," *Financial Times*, April 9, 2013.

43. Barry Eichengreen and Poonam Gupta, "The Service Sector as India's Road to Economic Growth," National Bureau of Economic Research, February 2011.

44. Deepak Lal, "An Indian Economic Miracle?" *Cato Journal* 28 (Winter 2008): 11–34.

45. Ruchir Sharma, "The Rise of the Rest of India: How States Have Become the Engines of Growth," *Foreign Affairs* 92 (September–October 2013): 75–85; James Crabtree, "Beware India's Moment of Modi Mania," *Financial Times*, December 10, 2013.

46. Barry Eichengreen, Donghyun Park, and Kwanho Shin, "When Fast Growing Economies Slow Down: International Evidence and Implications for China," National Bureau of Economic Research, March 2011; Barry Eichengreen, Donghyun Park, and Kwanho Shin, "Growth Slowdowns Redux: New Evidence on the Middle-Income Trap," National Bureau of Economic Research, January 2013.

47. Joseph A. Schumpeter, *The Theory of Economic Development* (Cambridge, MA: Harvard University Press, 1934), chap. 2; Joseph A. Schumpeter, *Capitalism, Socialism and Democracy* (New York: Harper and Row, 1942), chap. 7.

48. Robert J. Gordon, "Does the New Economy Measure Up to the Great Inventions of the Past?" *Journal of Economic Perspectives* 4 (Fall 2000): 49–74; Tyler Cowen, *The Great Stagnation* (New York: Dutton, 2011).

49. Barry D. Watts, *The Maturing Revolution in Military Affairs* (Washington, DC: Center for Strategic and Budgetary Assessments, 2011); Thomas G. Mahnken, "Weapons: The Growth and Spread of the Precision-Strike Regime," *Daedalus* 140 (Summer 2011): 45–57.

50. Bradford A. Lee, "Strategy, Arms and the Collapse of France, 1930–40," in *Diplomacy and Intelligence During the Second World War*, ed. Richard Langhorne (Cambridge: Cambridge University Press, 1985), 55–63.

12 NUCLEAR DETERRENCE IN NORTHEAST ASIA

Michael S. Chase

CONTEMPORARY SECURITY issues in East Asia tend to focus on developments involving conventional military capabilities, but the United States also faces challenges relating to nuclear weapons and proliferation. Such developments have serious implications for regional stability and the interests of the United States and its allies and partners. Some observers are relatively optimistic about the prospects for nuclear stability, while others will argue that the applicability of Cold War theories to deterrence is limited by the emergence of multiple nuclear powers and developments in areas of missile defense and conventional precision-strike capabilities.

The concerns over nuclear developments in China and North Korea pose questions about the future of East Asian security. China is modernizing its nuclear forces with road-mobile intercontinental ballistic missiles (ICBMs), nuclear-powered ballistic-missile submarines (SSBNs), and submarine-launched ballistic missiles (SLBMs). The focus of Beijing on an assured retaliation capability supports its strategy of deterrence. China's second-strike capability may promote stable mutual deterrence relations with the United States, but any crisis or conflict could entail miscalculation or miscommunication that triggers escalation.

The determination of North Korea to maintain and strengthen its nuclear capability poses different challenges. Beyond concerns over deterrence and nonproliferation, the United States faces the possibility that increased North Korean nuclear capabilities could lead to provocative actions, which could escalate if Pyongyang calculates that nuclear threats or even the use of nuclear weapons would improve its chances of survival. The remainder of this chapter consists of three parts: the lessons of deterrence theory and the Cold War nuclear strategic debates, an overview of challenges associated with Chinese and North Korean nuclear capabilities, and the implications of those developments for the United States.

DETERRENCE THEORY AND CONTEMPORARY CHALLENGES

Deterrence has a long history in international relations, but its modern usage is often traced to the airpower theorists of the interwar period, who argued that the only likely way to prevent massive destruction by enemy bombers would be to threaten would-be attackers with punishing retaliatory strikes. For many observers, however, the theory and practice of deterrence is most closely associated with Cold War foreign policy and nuclear strategy. For some, nuclear weapons ensured that the military competition between the United States and the Soviet Union did not lead to another great-power war. For others, nuclear weapons heightened the dangers of superpower confrontation.

Throughout the Cold War, analysts debated the implications of nuclear weapons. Setting the tone for the ensuing debate was Bernard Brodie, who was perhaps best known for the conclusion that in the nuclear age the primary purpose of military power would be deterring wars rather than winning them.[1] As the Cold War developed, scholars built on these earlier studies, and several debates emerged, one of which concerned the influence of nuclear weapons. On one side, some highlighted the far-reaching implications of these weapons. Robert Jervis concluded that nuclear warfare, especially with secure second-strike capabilities and mutually assured vulnerability, led to a revolution in statecraft. For others, however, nuclear weapons were far from revolutionary, and winning a major-power war remained possible even in the nuclear era. But the definitions of victory that they used often failed to relate the potential cost of war to the value of the objectives at stake in a conflict. As Jervis points out, their conclusions were consistent with the definitions, which made their analysis remarkably apolitical.[2]

Many things might have changed with the advent of nuclear weapons, and perhaps equally important changes occurred after the Cold War, but one thing that remains unchanged is that deterrence involves influencing the decisions of other actors. As Lawrence Freedman writes, deterrence involves "the potential or actual application of force to influence the action of a voluntary agent." But whereas deterrence attempts to persuade another party not to act out of fear of the consequences, "compellence" seeks to convince them "that they must act for fear of the consequences if they do not."[3]

It is useful to contrast strategies of deterrence and compellence with *controlling* strategies. The point of deterrence and compellence is influencing decisions rather than eliminating the capacity to make decisions.[4] Most scholars agree that, as a result, deterrence requires three things: capability, will, and communication. First, actors must have adequate capabilities. Second, they must have the resolve to use the capabilities at their disposal. Third, they must communicate that they possess the required capabilities and will employ them if necessary.

Of many distinctions in deterrence theory, such as between counterforce and countervalue targeting or between mutually assured destruction and damage limitation, one of the most noteworthy is between deterrence by denial and deterrence by threat of punishment.[5] Deterrence by denial can influence an enemy by demonstrating sufficient defensive capabilities to render hostile actions ineffective, denying an enemy the ability to achieve the desired objectives. Deterrence by threat of punishment depends on the ability to inflict costs in response to the initiation of enemy action one seeks to deter; in other words, it may not be possible to prevent an enemy from achieving its objectives, but it can be demonstrated that the punishment to be inflicted in response will impose costs that outweigh the value of whatever may be gained. Because adversary policy makers weigh the costs and benefits of action against those of inaction, another crucial aspect of deterrence is encouraging restraint by convincing enemy policy makers that refraining from taking the action that one seeks to deter will result in an acceptable outcome.[6]

Deterrence theory has been subject to criticism, including that it is notoriously difficult to prove its success in a given case. In addition, it is premised on the bipolar international system that prevailed during the Cold War, which potentially limits its utility today. In practical terms, another concern was that even under relatively favorable circumstances, deterrence failures can result from miscommunication, lack of credibility, asymmetrical interests, and perceived costs of not taking action. Scholars also raised concern over an inadvertent escalation of conventional conflicts among nuclear powers.[7] Consequently, deterrence fell out of fashion after the end of the Cold War and the collapse of the Soviet Union.

Some two decades later many analysts argue that Cold War deterrence theory has limited relevance. They articulate a number of reasons: a different international strategic context; questions about whether certain adversaries can be deterred in the traditional sense of the word; and advances in conventional capabilities, such as precision strike, information warfare, and space operations, which may have strategic effects even though they do not cross the nuclear threshold.

The debate about the contemporary significance of deterrence theory is clearly related to understanding emerging security dynamics in Asia. Some observers are relatively optimistic about the future of nuclear weapons in the region.[8] But others are less sanguine about nuclear developments and the prospects for regional stability. Indeed, studies of the changing nuclear dynamics in Asia have posited serious concerns about strategic competition in the nuclear arena and its likely meaning for regional security, escalation risks, and US and allied vital interests. For one, the rivalry between nuclear-armed superpowers during the Cold War proved to be extremely dangerous at times, even if it contributed to a long period when there were no conflicts among the major powers. In addition, circumstances that characterized that period do not apply to the contemporary Asian security context.

As Christopher Twomey observes, the current situation in Asia "adds several layers of exacerbating factors" to the dangers of the Cold War. In particular, he highlights three principal changes: the emergence of a multipolar nuclear environment, the development of systems other than nuclear weapons that have the potential to create strategic effects, and the different views that relevant nations hold about nuclear weapons. First, the multipolar nuclear environment adds several players and greater complexity to the relatively simple bipolar structure of the Cold War, creating a larger number of relationships among Asian nuclear powers. Second, the development and diffusion of conventional precision-strike capabilities and missile defense systems could be destabilizing. Third, different views on nuclear weapons as instruments of political and military power could add strategic complexity to an already volatile mix, especially in concert with the two other changes. According to Twomey, "While each change leads to dangerous outcomes in isolation, their combined effects are even more deleterious."[9]

The relatively pessimistic assessment expressed by Twomey is based on examining Asia in general, while the focus here is limited to East Asia, specifically China and North Korea. The rationale for the selection of East Asia is that it is the part of Asia in which the possibility of a crisis or conflict involving direct US military intervention against another nuclear power appears most plausible. In South Asia the major challenges faced by the United States are securing Pakistani weapons, nonproliferation, and preventing nuclear escalation between India and Pakistan.[10] Russia remains the only other nuclear superpower and an important potential security challenge for the United States, but the major security challenges in the US-Russia relationship are related to European security issues. In contrast, unfortunately, it is not difficult to construct a scenario in which the United States could become embroiled in a crisis or conflict in East Asia, perhaps one that results from North Korean provocations of South Korea or the Chinese use of force against Taiwan. Other possibilities include a Sino-Japanese conflict over the Senkaku/Diaoyu Islands or war in the South China Sea between China and Vietnam or the Philippines. Any scenario involving a nuclear-armed enemy, whether a small, regional nuclear power or a major power, would seriously challenge the United States.

POTENTIAL NUCLEAR ADVERSARIES

The United States must contend with two nuclear powers in East Asia: China and North Korea. China has been a nuclear power since 1964. Beijing is acquiring a secure assured-retaliation capability after relying on a vulnerable minimal-deterrence nuclear force for decades.[11] North Korea demonstrated its ability to produce a nuclear device in 2006, despite strong international opposition. Pyongyang is determined to maintain at least a small nuclear force for deterrence and bargaining leverage.[12] The two nations present different challenges for the United States.

China

The nuclear strategy of the People's Republic of China rests on assured-retaliation capabilities to deter a nuclear-armed enemy. Beijing conducted its first nuclear test in 1964 and has maintained a no-first-use policy for more than four decades. From the 1960s to early 1980s, China's potential was highly limited. China began fielding silo-based ICBMs in the 1980s but deployed a modest and possibly vulnerable nuclear force to support minimal deterrence. In addition, China has been modernizing theater and strategic nuclear forces to enhance survivability, expand striking power, and counter developments in missile defense. The transition to survivable nuclear forces such as mobile ICBMs and SSBNs is the most notable aspect of force modernization and is closely aligned with the nuclear strategy developed by Beijing.

China is moving from a traditional minimal deterrence strategy, under which ambiguity and the threat of retaliation with a handful of nuclear weapons were seen as sufficient for deterrence, to one that is sometimes characterized as assured retaliation, which depends on more forces surviving a first strike and penetrating enemy missile defenses. Some Chinese scholars have described this approach as "dynamic minimum deterrence."[13] They argue that it retains features of traditional Chinese strategy while keeping pace with changes in the security environment.

With regard to nuclear policy and strategy, a Chinese defense white paper issued in 2006 reiterated that Beijing unconditionally undertook not to use or threaten the use of nuclear weapons against nonnuclear states and promoted the comprehensive prohibition and elimination of such weapons.[14] Other authoritative sources on missile forces and deterrence shed further light on Chinese nuclear strategy. One question these publications address is what types of enemy actions Beijing thinks its nuclear capability can deter.

Chinese strategists expect deterrence to prevent not only enemy use of nuclear weapons but also certain types of conventional attacks. In the event of a conflict with a nuclear power, the Second Artillery will conduct nuclear deterrence operations to influence enemies and constrain their options. China has said that a conventional missile strike campaign would be carried out under "nuclear deterrence conditions."[15] This illustrates that the Chinese view nuclear weapons as a "backstop to support conventional operations."[16]

Nuclear deterrence thus plays an important role during conventional conflicts, but Chinese doctrinal publications like the *Science of Second Artillery Campaigns* continue to reflect the longstanding no-first-use policy in that they assume the Second Artillery nuclear forces would launch their weapons only after an enemy first strike. This is also consistent with statements by Chinese experts who assert that their nation will continue to adhere to a no-first-use policy. For instance, according to Rong Yu and Peng Guangqian, "Although some people believe China's no-first-use policy is not credible, China has never wavered from its promise during the past 40 years."[17] Nonetheless, Chinese strategists still debate nuclear policy and strategy, including what kinds of actions nuclear weapons may be able to deter apart from nuclear threats or attack and how best to respond to the challenges posed by advances in enemy intelligence, surveillance, and reconnaissance; precision-strike; and missile defense capabilities.

Chinese strategists also suggest that there is some ambiguity when it comes to determining what constitutes first use by an enemy. As Rong and Peng have stated, because conventional attacks may have devastating effects in certain instances, "definitely establishing whether the adversary has broken the nuclear threshold is not necessarily a straightforward issue." They raise the specific question of whether a conventional attack on a nation's nuclear forces could be considered tantamount to nuclear first use. "On the surface, this is merely a conventional attack," they write, "but in effect, its impact is little different than suffering a nuclear strike and incurring similarly heavy losses."[18] Thus, conventional attacks might cross the nuclear threshold, with the targets of such attacks unable to refrain from counterattacking with nuclear weapons.

Other analysts have considered scenarios under which China might consider departing from its traditional no-first-use policy in response to enemy conventional attacks on targets such as its nuclear forces, major urban areas, nuclear power plants, and key hydroelectrical facilities (perhaps a subtle reference to the Three Gorges Dam). One scenario involves threatening the territorial integrity of China in a conventional conflict presumably because of the no-first-use policy. Another scenario includes the possibility of nuclear retaliation for conventional strikes on strategic sites in an intervention by a major power such as the United States. Nonetheless, Chinese scholars suggest that Beijing would approach any decision to authorize the use of nuclear weapons with caution and that such decision would be made only under extreme circumstances. Given the damage that would be caused, the decision to use nuclear weapons would be "imaginable [only] if core national interests are in peril, such as the survival of the state or nation."[19]

China maintains limited but growing nuclear forces consistent with its goal of sustaining an assured retaliation posture in support of its nuclear deterrence strategy. This includes forces aimed at regional targets as well as ICBMs. For regional nuclear deterrence, China fields DF-3 intermediate-range ballistic missiles (IRBMs; liquid-propellant missiles, which were first deployed in 1971) and the more modern and survivable road-mobile DF-21 and DF-21A medium-range ballistic missiles (MRBMs).[20] As for Chinese longer-range forces, the silo-based DF-5 ICBM, a liquid-propellant, silo-based missile, has been the mainstay of its deterrence force since being deployed in 1981. China today deploys about twenty DF-5s, which are capable of striking targets in the continental United States.[21] Converting DF-5s to a multiple independently targetable reentry vehicle configuration would dramatically increase the number of warheads that China could launch.[22]

In addition, the Chinese deploy DF-31s and DF-31As, which are road-mobile ICBMs. The DF-31 was fielded in 2006 after a long period of development that began in the 1980s. It is probably intended to cover targets in Asia and Russia, but its range is sufficient to reach missile defense sites in Alaska, US forces in the Pacific, and the western regions of the United States.[23] China has fielded five to ten DF-31 ICBM launchers. DF-31As, which China began to deploy in 2007, are three-stage ICBMs that can reach targets in the continental United States. More than fifteen DF-31A ICBM launchers were fielded as of 2013.[24] In addition, China maintains some older limited-range, liquid-fueled DF-4 ICBMs that were first deployed in 1975.

China is expected to field SLBMs to complement its land-based nuclear ballistic missiles. Indeed, the Chinese undersea deterrent is undergoing a major change with the emergence of the type 094, or *Jin*-class, nuclear-powered SSBNs, which represent a substantial improvement over first-generation type 092, or *Xia*-class, SSBNs. The type 092 was launched in the early 1980s but has never conducted a deterrent patrol.[25] In contrast, China appears determined to build enough type 094 SSBNs to conduct near-continuous deterrent patrols.[26] A 2013 Department of Defense report on Chinese military and security developments notes that China has delivered three *Jin*-class SSBNs to the People's Liberation Army Navy, with "as many as two more in various stages of construction."[27] Although China has faced some problems with the development of the JL-2 SLBM, according to the report, "after a round of successful testing in 2012, the JL-2 appears ready to reach initial operational capability in 2013."[28]

The number of Chinese nuclear weapons capable of striking targets in the United States is likely to increase and will probably exceed one hundred as their forces are modernized.[29] In addition, China may develop one or more newer ICBMs. According to a report issued by the Pentagon in 2010, "China may also be developing a new road-mobile ICBM, possibly capable of carrying a multiple independently targeted re-entry vehicle."[30] Chinese thinking about nuclear weapons argues for a larger and more diverse force to ensure an assured retaliation capability that remains credible despite advances in intelligence, surveillance, and reconnaissance; strike; and missile defense capabilities. It also suggests that Beijing would not perceive dramatic benefits from matching Russia or the United States in overall nuclear force structure.

In addition to nuclear-capable missile systems, China has conventional ballistic missiles that can deliver precision or near-precision strikes. They include short-range ballistic missiles (SRBMs), including both older systems that lack precision-strike capability and newer systems that have improved range, accuracy, and payloads such as unitary and submunition warheads; MRBMs, which increase the range of strikes on land and sea targets, including in the first island chain; land-attack cruise missiles (LACMs), which are being developed as air- and ground-launched systems for standoff precision strikes; ground attack munitions, which include tactical air-to-surface missiles and precision-guided munitions (all-weather, satellite-guided bombs, antiradiation missiles, and laser-guided bombs); antiship cruise missiles (ASCMs), which are fielded in many variants, including Russian SS-N-22 and SS-N-27B; antiradiation weapons, which include both Israeli Harpy unmanned combat aerial vehicles and Russian antiradiation missiles; and artillery-delivered high-precision munitions, which are able to strike targets within or even across the Taiwan Strait.[31] Importantly, the Second Artillery controls conventional ballistic missiles and ground-based LACMs as well as most nuclear missiles. Because one component of the military is responsible for conventional missile strikes and nuclear deterrence, risks of inadvertent escalation in a crisis or conflict cannot be discounted.[32]

North Korea

The nuclear weapons programs and proliferation activities of the Democratic People's Republic of Korea (DPRK) have challenged the United States and its allies for many years. US policy responses to these challenges have included sanctions, strengthened alliances, nonproliferation measures, and two major diplomatic initiatives aimed at convincing North Korea to abandon its nuclear activities. The first initiative took place in 1994 when the United States negotiated the Agreed Framework, under which North Korea was to freeze its nuclear weapons program in exchange for heavy fuel oil and construction of two proliferation-resistant light-water reactors. But in 2002 that accord fell apart following North Korea's acknowledgment of a uranium enrichment program for nuclear weapons, and the next year Pyongyang withdrew from the Non-Proliferation Treaty and continued to pursue nuclear capabilities. The second initiative was the Six-Party Talks that began in 2003 and involved the two Koreas, China, Japan, Russia, and the United States. The talks seemingly achieved a breakthrough, when Pyongyang promised to give up nuclear programs, but the process collapsed in 2009, and North Korea repeated its intention to develop nuclear weapons, which it regarded as essential to its security and strategic autonomy.[33] Pyongyang probably also views its program as successful in garnering economic assistance and diplomatic attention that helped to ensure its survival.

Today, North Korean nuclear and missile programs remain major challenges to the United States and its allies. Proliferation activities by North Korea, including exporting ballistic missiles to Iran and Syria and involvement in construction of the Syrian nuclear reactor destroyed by Israel, also loom as threats to security and stability in the region and throughout the world.

Official statements by North Korea have provided a rough outline of its policy on nuclear weapons. For example, a report by the Korean Central News Agency summarized the thinking of Pyongyang on deterrence: "The mission of the nuclear armed forces of the DPRK is to deter and repulse aggression and attack on the country and the nation till the nuclear weapons are eliminated from the peninsula and the rest of the world."[34] It is clear that the motives for developing nuclear and missile capabilities include compensating for weak conventional assets with nuclear deterrence, enhanced international and domestic prestige of its leader, and greater leverage to extract diplomatic and economic concessions.

Less is known about circumstances under which Pyongyang would use nuclear weapons against the United States and its allies. According to James Clapper, director of National Intelligence, North Korea "would consider using nuclear weapons only under certain narrow circumstances." Specifically, Clapper said that intelligence assessments have concluded, "albeit with low confidence[, that Pyongyang] probably would not attempt to use nuclear weapons against US forces or territory unless it perceived its regime to be on the verge of military defeat and risked an irretrievable loss of control."[35]

The Defense Intelligence Agency (DIA) has determined that the North Koreans could have several nuclear warheads that could be delivered by ballistic missiles, aircraft, or unconventional means.[36] But the extent of their nuclear weapons capability is unclear. "We do not know whether the North has produced nuclear weapons, but we assess it has the capability to do so," Clapper said.[37]

Although the specifics of its weapons capabilities are unknown, North Korea has conducted two underground nuclear tests. Its first test of a nuclear device occurred October 9, 2006. The Office of the Director of National

Intelligence released a statement roughly one week later with an overview of the US assessment of this detonation. According to the press release, "Analysis of air samples collected on October 11, 2006, detected radioactive debris which confirms that North Korea conducted an underground nuclear explosion in the vicinity of P'unggye on October 9, 2006." But the yield of this test was less than a kiloton.[38] US intelligence sources determined the test was consistent with the assessment that North Korean had produced a nuclear device but judged it to be a partial failure.[39] Another test was conducted May 25, 2009.[40] That test was somewhat larger and equivalent to approximately two kilotons of chemical high explosives.[41] Overall, the latter test appeared more successful, confirming a judgment that North Korea was making progress in its research and development of nuclear weapons.[42] North Korea conducted its third nuclear test on February 12, 2013, and in March 2014 Pyongyang suggested that it might conduct an unspecified "new form" of nuclear test.[43]

According to a report by the National Air and Space Intelligence Center, "North Korea has an ambitious ballistic missile development program and has exported missile technology to other countries, including Iran and Pakistan."[44] Pyongyang "is continuing to develop the [Taepo-Dong 2, which] could reach the United States with a nuclear payload if developed as an ICBM. An intermediate-range ballistic missile (IRBM) and a new short-range, solid-propellant ballistic missile are also being developed."[45] Finally, North Korea unveiled a new road-mobile ICBM in an April 2012 military parade.

As for short-range missile forces, North Korea deploys several different types of SRBMs. As of 2013, it had fewer than one hundred SRBM launchers in all.[46] It also has a mobile medium-range missile called the No-Dong and is developing IRBMs. Pyongyang has used these capabilities in several high-profile missile tests since the 1990s. It conducted the first Taepo-Dong 1 launch in 1998. The United States found no evidence to confirm the North Korean claim that the Taepo-Dong 1 put a satellite in orbit, but the test "demonstrated technologies necessary for longer-range missile development."[47] Perhaps more important was that the missile overflew Japan before landing in the Pacific, fueling concern about North Korean missile and nuclear capabilities.

A Taepo-Dong 2 test in July 2006 failed shortly after takeoff, but it was accompanied by the successful launch of a number of SRBMs and MRBMs. In April 2009 Pyongyang again launched the long-range Taepo-Dong 2, which may be capable of reaching parts of the United States. The US intelligence community believes the April 2009 launch of the Taepo-Dong 2 "ended in failure."[48] North Korea claims that the 2009 launch placed a communications satellite into orbit. Although the launch apparently failed to achieve this objective, it "successfully tested many technologies associated with an ICBM." It also "demonstrated a more complete performance than the July 2006 launch."[49] The launch underscored Pyongyang's determination to develop space-launch and ICBM capabilities despite international objections.[50] Another launch failed in April 2012, but in December of that year, North Korea finally placed a satellite into orbit using a Taepo-Dong 2. In addition, Pyongyang is developing the new Hwasong-13 road-mobile ICBM.[51]

Pyongyang has indicated its intention of returning to the Six-Party Talks, but according to the Defense Intelligence Agency, "Pyongyang is unlikely to eliminate its nuclear weapons."[52] After NATO air strikes on Libya in 2011, the Korean Central News Agency issued a statement by the Foreign Ministry declaring that the Libyan decision to relinquish nuclear weapons exposed it to Western intervention. According to the statement, the Libyan nuclear dismantlement "turned out to be a mode of aggression whereby the [United States] coaxed [Libya] with such sweet words as 'guarantee of security' and 'improvement of relations' to disarm itself and then swallowed it up by force."[53] This statement appeared intended to underscore the importance that North Korea attached to its nuclear capabilities as a deterrent to the United States.

One reason for this determination to hold on to nuclear capabilities rather than trade them for Western diplomatic and economic concessions could be the shortcomings of North Korean conventional capabilities. According to Clapper, "Although there are other reasons for the North to pursue its nuclear program, redressing conventional weakness is a major factor and one that Kim [Jong-il] and his likely successors will not easily dismiss."[54] It is not a question of North Korea denuclearizing but rather whether it will maintain a symbolic capability or develop an operational deterrent.[55] The apparent pursuit by Pyongyang of a road-mobile ICBM capability indicates that it seeks the latter.

IMPLICATIONS FOR THE UNITED STATES

Recent developments in Chinese and North Korean nuclear strategy and capabilities suggest that the United States will need to deal with two different nuclear strategy and deterrence challenges: maintaining a stable relationship with China on the strategic level and coping with the effects of the North Korean determination to be known as a nuclear power. The challenges posed by these developments could have a serious impact on the United States and its allies.

Maintaining Strategic Stability

Chinese nuclear modernization is focused on improving the capability to survive a first strike by an enemy and making its nuclear deterrence posture more credible—objectives that Chinese analysts believe have gained increased urgency as a result of growing US nuclear preeminence, missile defense plans, and conventional precision-strike capabilities. China is moving toward more survivable and more credible strategic nuclear capabilities with the deployment of both road-mobile DF-31 and DF-31A ICBMs and JL-2 SLBMs. Indeed, road-mobile strategic missiles as well as SSBNs are enabling the Chinese military to achieve the "lean and effective" deterrence posture called for in its official publications.[56]

The modernization of Chinese nuclear forces and the transition from silo-based to road-mobile nuclear missiles and SSBNs might thus enhance strategic deterrence stability. Indeed, deterrence theory suggests that a more secure second-strike capability would enhance stability by causing both the United States and China to behave much more cautiously. Yet there are a number of reasons to be concerned that the transition to a more secure second-strike capability will not offer greater stability in the event of a US-China crisis like the Sino-Japanese dispute over the Senkaku/Diaoyu Islands, South China Sea, or Taiwan Strait.

Miscalculation in the midst of a crisis is a particularly troubling possibility, one that could be heightened by uncertainty over the message that one side is attempting to convey to the other or overconfidence in the ability to control escalation. Some of the signaling activities described in Chinese publications could easily be misinterpreted as preparation to conduct nuclear missile strikes, possibly decreasing crisis stability or even triggering escalation rather than strengthening deterrence. Indeed, some Chinese sources cite troubling issues on potential miscalculations that could result from attempts to enhance deterrence in crises or during conventional conflicts. For example, one method of deterrence that has been suggested in Chinese publications involves the simulated replenishment of liquid-fueled missiles.[57]

Although this tactic is intended to place an enemy under severe pressure by indicating that missile forces have entered a premobilization state, the Chinese apparently fail to consider the potential for miscalculation. Given the risk of unintended escalation, this stratagem could result in a destabilizing action, especially during a conflict with a nuclear-armed adversary. Although some Chinese sources acknowledge that the risk of actions intended to deter an enemy could trigger escalation, the arguments are not well developed.[58] Another problem for the United States is that the development of the wherewithal that would be required in a confrontation with North Korea, such as missile defense and conventional global-strike capabilities, could lead to unintended consequences. Those assets could exacerbate tensions with China if Beijing sees them as a threat to its security, and they could possibly undermine other US policy objectives, most notably in promoting strategic stability with other major nuclear powers as outlined in the 2010 "Nuclear Posture Review Report."[59]

China has argued that US missile defense systems and proposed conventional global-strike programs would negatively impact strategic stability by compromising its assured second-strike capability. Specifically, Beijing has suggested that such capabilities would make it easier for Washington to contemplate a first strike. Indeed, Chinese analysts view the pursuit of a missile defense system by the United States as a significant threat to the viability of nuclear deterrence. Moreover, some Chinese scholars have expressed concerns that even if US missile defense and conventional global-strike systems have little or no real impact on a second strike, US planners and policy makers may develop a false sense of superiority, potentially leading to attempts to coerce China with nuclear threats in a crisis. Similarly, they suggest that the illusion of nuclear primacy could lead to more aggressive behavior by the United States.[60]

Beijing is concerned that such capabilities could undermine strategic stability. As a result, Washington must act carefully to avoid precipitating responses that are contrary to its interests, such as a larger than planned Chinese nuclear force buildup, further advances in counter–space capabilities, or potentially destabilizing higher alert levels. Consequently, as China modernizes its nuclear and missile forces, the problems of strategic stability appear poised to become much more critical aspects of the US-China security relationship in the future. Thus, although Chinese nuclear and missile force modernization may contribute to greater strategic stability in the long run, neither China nor the United States should assume that this outcome will arise automatically from the Chinese transition to a secure second-strike capability.

These emerging dynamics of the US-China strategic relationship thus underscore the need to further enhance the bilateral dialogue on strategic stability and nuclear weapons issues. At least some Chinese experts and military personnel appear to recognize the potential value of expanding consultations on strategic issues. At the same time, however, the persistent concerns raised by Beijing over potential risks of greater transparency may limit its willingness to engage with Washington. For its part, the United States should concentrate on enhancing dialogue by persuading China that increasing transparency would not undermine its interests but would benefit both sides by helping promote shared interests in strategic stability. Nonetheless, the United States could find it difficult to overcome long-standing concerns on the part of China. Indeed, some Chinese experts have warned that uncertainty over technological developments may result in even greater anxiety about nuclear transparency.

Coping with a Nuclear North Korea

North Korea appears determined to maintain and perhaps further develop its nuclear capabilities. And as Jonathan Pollack observes, there is almost no chance that it will relinquish its nuclear ambitions. Instead, events in the past indicate that Kim Jong-il "unambiguously tethered North Korea's long-term security to nuclear weapons development, not to any presumed benefits that denuclearization might provide."[61] Pyongyang's determination to retain and possibly expand its nuclear capabilities under Kim Jong-un poses several potentially serious challenges.

First, planners and policy makers must consider the implications of North Korea's nuclear capabilities in the event of a conventional conflict on the Korean Peninsula. In particular, they must consider the possibility that North Korea would use nuclear weapons if its leaders thought it was necessary to avoid a conventional defeat that could threaten the survival of the regime. As some observers point out, leaders of countries that would face such a defeat in a conflict with conventionally superior US forces might consider using nuclear weapons to compel the United States to cease conducting military operations.[62] Moreover, fear of US attacks against leadership targets, command-and-control systems, and nuclear capabilities and delivery systems could intensify pressure on an enemy to use nuclear weapons before losing that ability. Thus, deterrence by threat of punishment may not be a viable strategy under such circumstances, leaving the United States with three options: developing capabilities to prevent a regional power like North Korea from using nuclear weapons, limiting military operations to minimize the risk of triggering nuclear escalation, or avoiding conflict in the first place.

Another concern is the possibility that nuclear capabilities could embolden Pyongyang to take greater risks and engage in more-dangerous provocations. In addition, the conventional threat posed by North Korea to South Korea is a serious deterrent to retaliation. This presents major challenges in dealing with North Korea because it makes provocations more difficult to deter than an invasion, which demands the attention of both Washington and Seoul.[63] The problem of deterring provocations underscores the point that North Korean attacks on South Korea may create the greatest danger of escalation, and the transition from Kim Jong-il to Kim Jong-un appears to have increased that risk. Indeed, the leadership succession seems to be connected with recent provocations. The new leadership in Pyongyang may expand use of provocations to achieve broad policy objectives through coercive diplomacy. Indeed, the recent provocations appear to have been intended to strengthen the position of Kim Jong-un as successor to his father. Domestic dynamics in North Korea eventually may create incentives for additional attacks.

The North may calculate that it can get away with further provocations because its nuclear capabilities and the risk of escalation decrease the likelihood of a firm response by South Korea or the United States. Although North Korea seems to be confident that its provocations will not trigger escalation, this could be a grave miscalculation. Indeed, after the sinking of the *Cheonan* in 2010 and especially the shelling of Yeonpyeong Island some months later and the extremely angry reaction by the South Korean public, a further attack would put tremendous pressure on Seoul to respond decisively. Indeed, after the incidents, President Lee Myung-Bak promised a powerful counterattack in response to aggression.[64]

Another risk is the possibility that North Korean nuclear weapons could give Japan and South Korea incentive to pursue their own nuclear capabilities. Moreover, North Korea could export nuclear technology to nations like Syria and Iran. The challenges for the United States also include assuring its allies that they can depend on extended deterrence and prevention of Pyongyang from proliferating nuclear technology.

• • •

One major challenge faced by the United States may be ensuring that responses to issues posed by both Chinese and North Korean nuclear and ballistic missile capabilities do not work at cross-purposes. It must be careful that actions taken with regard to North Korean nuclear programs do not inadvertently undermine efforts to forge stable strategic relations with China. As Pyongyang continues developing those programs, it may become more difficult to take steps to counter them without increasing Chinese suspicions of US strategic intentions. At the same time, Washington may gain leverage from its responses to Pyongyang. The desire by China to avoid actions by the United States that it sees as undermining its security interests—such as increased missile defense interceptors to counter the mounting North Korean threat and naval exercises with Japan and South Korea near the Chinese mainland—may provide Beijing with a clear incentive to support attempts by Washington to put greater pressure on Pyongyang.

NOTES

1. Frederick S. Dunn, Bernard Brodie, Arnold Wolfers, Percy E. Corbett, and William T. R. Fox, eds., *The Absolute Weapon: Atomic Power and World Order* (New York: Harcourt Brace, 1946).

2. Robert Jervis, *The Meaning of the Nuclear Revolution* (Ithaca, NY: Cornell University Press, 1984), 19.

3. Lawrence Freedman, *Deterrence* (Cambridge, UK: Polity, 2004), 26–27.

4. Ibid., 85.

5. Glenn Snyder, *Deterrence and Defense* (Princeton, NJ: Princeton University Press, 1961).

6. United States Department of Defense, "Deterrence Operations Joint Operating Concept, Version 2.0," December 2006.

7. Barry Posen, *Inadvertent Escalation: Conventional War and Nuclear Risks* (Ithaca, NY: Cornell University Press, 1991).

8. For a relatively optimistic take, see Muthiah Alagappa, ed., *The Long Shadow: Nuclear Weapons and Security in 21st Century Asia* (Stanford, CA: Stanford University Press, 2008).

9. Christopher P. Twomey, "Asia's Complex Strategic Environment: Nuclear Multipolarity and Other Dangers," *Asia Policy* 11 (January 2011): 53.

10. For more on discouraging two states from going to war, see Timothy W. Crawford, *Pivotal Deterrence: Third Party Statecraft and the Pursuit of Peace* (Ithaca, NY: Cornell University Press, 2003).

11. M. Taylor Fravel and Evan S. Medeiros, "China's Search for Assured Retaliation: The Evolution of Chinese Nuclear Strategy and Force Structure," *International Security* 35 (Fall 2010): 48–87.

12. John S. Park and Dong Sun Lee, "North Korea: Existential Deterrence and Diplomatic Leverage," in Alagappa, *The Long Shadow*, 269–95.

13. Chu Shulong and Rong Yu, "China: Dynamic Minimum Deterrence," in Alagappa, *The Long Shadow*, 161–87.

14. People's Republic of China, "China's National Defense in 2006" (white paper, Information Office of the State Council, Beijing, December 29, 2006).

15. Xijun Zhao, ed., *Intimidation Warfare: A Comprehensive Discussion on Missile Deterrence* (Beijing: National Defense University Press, 2005), 88, 93.

16. People's Liberation Army, Second Artillery Force, *Science of Second Artillery Campaigns* [第二炮兵战役学, *Dier paobing zhanyixue*] (Beijing: PLA Press, 2004), 273.

17. Rong Yu and Peng Guangqian, "Nuclear No-First-Use Revisited," *China Security* 5 (Winter 2009): 89.

18. Ibid., 85.

19. Ibid., 88.

20. United States Department of Defense, National Air and Space Intelligence Center (NASIC), *Ballistic and Cruise Missile Threat* (Wright-Patterson Air Force Base, OH: National Air and Space Intelligence Center, 2009), 14; United States Department of Defense, *Annual Report to Congress: Military Power of the People's Republic of China, 2009* (Washington, DC: Office of the Secretary of Defense, 2009), 66.

21. NASIC, *Ballistic and Cruise Missile Threat* (2009), 21.

22. Brad Roberts, "Nuclear Minimalism," *Arms Control Today*, May 2007, https://www.armscontrol.org/act/2007_05/BookReview.

23. Robert D. Walpole, "The Ballistic Missile Threat to the United States," Senate Subcommittee on International Security, Proliferation, and Federal Services, February 9, 2000.

24. National Air and Space Intelligence Center (NASIC), *Ballistic and Cruise Missile Threat*, NASIC-1031-0985-13 (Dayton, OH: Wright-Patterson Air Force Base, 2013), 21.

25. NASIC, *Ballistic and Cruise Missile Threat* (2013), 25.

26. United States Department of the Navy, "Seapower Questions on the Chinese Submarine Force," December 20, 2006.

27. United States Department of Defense, *Annual Report to Congress: Military and Security Developments Involving the People's Republic of China, 2013* (Washington, DC: Office of the Secretary of Defense, 2013), 31.

28. Ibid.

29. NASIC, *Ballistic and Cruise Missile Threat* (2009), 3.

30. United States Department of Defense, *Annual Report to Congress: Developments Involving the People's Republic of China, 2010* (Washington, DC: Office of the Secretary of Defense, 2010), 2.

31. Ibid., 31.

32. John W. Lewis and Xue Litai, "Making China's Nuclear War Plan," *Bulletin of the Atomic Scientists*, September 14, 2012, pp. 45–65.

33. Jonathan Pollack, *No Exit: North Korea, Nuclear Weapons, and International Security* (New York: Routledge, 2011).

34. Democratic People's Republic of Korea, "Foreign Ministry Issues Memorandum on N-Issue," Korean Central News Agency (Pyongyang), April 21, 2010.

35. James R. Clapper, "Worldwide Threat Assessment of the US Intelligence Community," United States Senate Committee on Armed Services, March 10, 2011, p. 6.

36. Ronald L. Burgess, "Worldwide Threat Assessment," United States Senate Committee on Armed Services, March 10, 2011.

37. Clapper, "Worldwide Threat Assessment," 5.

38. "Statement by the Office of the Director of National Intelligence on the North Korea Nuclear Test," news release 19-06, Office of the Director of National Intelligence, October 16, 2006.

39. Clapper, "Worldwide Threat Assessment," 5.

40. "Statement by the Office of the Director of National Intelligence on North Korea's Declared Nuclear Test on May 25, 2009," news release 23-09, Office of the Director of National Intelligence, June 15, 2009.

41. Clapper, "Worldwide Threat Assessment," 5.

42. Ibid., 5.

43. Choe Sang-Hun, "North Korea Vows to Use 'New Form' of Nuclear Test," *New York Times*, March 30, 2014.

44. NASIC, *Ballistic and Cruise Missile Threat* (2009), 15.

45. Ibid., 3.

46. NASIC, *Ballistic and Cruise Missile Threat* (2013), 13.

47. NASIC, *Ballistic and Cruise Missile Threat* (2009), 15.

48. Clapper, "Worldwide Threat Assessment," 6.

49. Ibid.

50. NASIC, *Ballistic and Cruise Missile Threat* (2013), 19.

51. Ibid.

52. Burgess, "Worldwide Threat Assessment," 14.

53. Democratic People's Republic of Korea, "Foreign Ministry Spokesman Denounces US Military Attack on Libya," Korean Central News Agency (Pyongyang), March 22, 2011.

54. Clapper, "Worldwide Threat Assessment," 11.

55. Jonathan Pollack, "North Korea's Nuclear Weapons Program to 2015: Three Scenarios," *Asia Policy* 3 (January 2007): 105–23.

56. See, for example, People's Republic of China, "China's National Defense in 2006."

57. Xijun, *Intimidation Warfare*, 185.

58. For example, Zhao cautions that if "inappropriate timing is chosen, then deterrence may cause the situation to deteriorate, even leading to the eruption and escalation of war." *Intimidation Warfare*, 172.

59. United States Department of Defense, "Nuclear Posture Review Report," April 2010.

60. Li Bin and Nie Hongyi, "An Investigation of China-US Strategic Stability," *World Economics and Politics* 2 (2008): 60–61.

61. Jonathan Pollack, "Kim Jong-il's Clenched Fist," *Washington Quarterly* 32 (October 2009): 153–73.

62. David Ochmanek and Lowell H. Schwartz, *The Challenge of Nuclear-Armed Regional Adversaries* (Santa Monica, CA: RAND, 2008).

63. Michael A. McDevitt, "Deterring North Korean Provocations," *Brookings Northeast Asia Commentary* 46 (February 2011), http://www.brookings.edu/research/papers/2011/02/north-korea-mcdevitt.

64. Chico Harlan, "South Korean President Faces Conflicting Pressures as He Toughens North Korea Response," *Washington Post*, December 28, 2010.

13 ARMS RACES AND LONG-TERM COMPETITION

Thomas G. Mahnken

IN LIGHT OF THE SCOPE and pace of Chinese military developments and the response to them, it has become commonplace to hear references to an emerging arms race in Asia. Desmond Ball, for instance, refers to "substantial evidence of action-reaction dynamics, of an emerging complex arms race in Northeast Asia."[1] Use of the term *arms race* implies that an action by one party will provoke a direct reaction by the other. That presupposes actors are clearly focused, aware of the strategic environment, and quick to respond in predictable ways to their competitors. By contrast, Edward Luttwak maintains that China is strategically autistic, more preoccupied with internal concerns than displays of its national power.[2] Under that analysis one could expect domestic issues to govern Beijing in modernizing its forces given the current international situation.

Although Sino-American arms competition does exist, it is not the sort of tightly coupled action-reaction process often identified with arms races. To appreciate that distinction one must review theories on arms races and apply them within the context of Asian security affairs to the sources of competition. Although external factors influence Chinese procurement, including actions by the United States, internal dynamics drive the form and timing of Beijing's responses. This circumstance became worryingly observable in the development of conventional precision-guided ballistic missiles, including antiship ballistic missiles (ASBMs).

ARMS RACE THEORY

Arms competitions have long existed but went largely unanalyzed until the Cold War. Indeed, during that period a sizable body of literature on arms races emerged, driven by interest in competition between the United States and Soviet Union. Samuel Huntington describes an arms race as "a progressive, competitive peacetime increase in armaments by two states or coalitions of states resulting from conflicting purposes or mutual fears."[3] Subsequently, Colin Gray came to define an arms race as "two or more parties perceiving themselves to be in an adversary relationship, who are increasing or improving their armaments at a rapid rate and restructuring their respective military postures with a general attention to the past, current, and anticipated military and political behavior of the other parties."[4] Accordingly, an arms race has four basic elements. First, it must involve at least two parties. Second, each party must develop its force structure with reference to the competition. Third, each must compete with the other in the quantity or quality of their respective militaries. Finally, interactions must lead to a rapid increase in the quantity or quality of weapons.

Arms competitions can also vary considerably in both scope and pace. With this fact in mind, Barry Buzan coined the term *arms dynamic* to refer to pressures that make nations buy military capabilities and enhance their quantity and quality over time.[5]

External Sources of Competition

One group of scholars emphasizes external sources of arms competition. The most common and simplistic formulation is the action-reaction model, which stipulates "that states strengthen their armaments because of the threats the states perceive from other states. The theory implicit in the model explains the arms dynamic as driven primarily by factors external to the state."[6] Basically, it holds that the search for security, together with uncertainty and worst-case estimates of enemy intentions and capabilities, will yield efforts to amass ever-greater stockpiles of weaponry. That means that exaggerated fears and inflated estimates of the threat will lead to spiraling growth of armaments and arms spending. This is supported by plans to field new weapons often being made before the systems they are intended to counter appear on the scene.[7]

Central to action-reaction arms race theory is the *security dilemma*. Robert Jervis observes that such a dilemma could be said to exist when "the means by which a state tries to increase its security decrease the security of others."[8] In other words, any nation that pursues security tends to conclude that enemy motives are more malign than generally thought and thus reacts accordingly. This dilemma is driven by uncertainty over whether the competitor is motivated by security or expansionist designs.[9] Jervis argues that the magnitude and nature of the security dilemma depend on the offense-defense balance as well as differentiation between offense and defense.

One major deficiency in the action-reaction notion of arms competitions is that it denies the role of policy and strategy. In fact, national objectives play a major role in determining the shape of armed forces and the scope and pace of modernization. A nation might acquire weapons because it fears attack or harbors expansionist designs. The German naval buildup at the start of the twentieth century, for example, was driven more by a craving for status and a greater international role than as a response to Great Britain's Royal Navy.[10] Although construction of warships by Great Britain and Germany was coupled, action-reaction theory fails to capture the dynamic.[11]

Action-reaction theory is based on the assumption that the decisions made by competitors on arms are closely coupled—that is, actors pay close attention to one another, and the magnitude and timing of their responses are directly related to actions by opponents. In practice, however, the awareness of such actions can vary considerably. Accordingly, differentiating between military modernization and arms races can be challenging.[12] In fact, the action-reaction model presents just one hypothesis on how and why external events lead to arms competition. The reasons for competition include deterring military action, achieving favorable outcomes in war, increasing diplomatic leverage, procuring arms because of competitors' modernization, and enhancing national standing. Finally, the emergence of new military technology that ensures block obsolescence of arms may drive competition.[13]

The action-reaction dynamic is appealing because of its simplicity. As Secretary of Defense Robert McNamara once remarked, "Whatever their intentions or our intentions, actions—or even realistically potential actions—on either side relating to the buildup of nuclear forces necessarily trigger reactions on the other side. It is precisely this action-reaction phenomenon that fuels the arms race."[14] Similarly, George Rathjens stresses that uncertainty over Soviet capabilities tended to cause an overreaction because the United States based its planning on worst-case assumptions. It also planned on the basis of Soviet capabilities that had not emerged.[15]

Despite its theoretical appeal, can the action-reaction dynamic explain past arms races, let alone serve as a useful means of interpreting present and future interaction? For example, though arms race theory developed during the Cold War, it is questionable whether it explained Soviet-American interaction. A study by Andrew Marshall of Soviet defense expenditures in the 1960s revealed a much looser interaction between Washington and Moscow than action-reaction theory predicted. Understanding that the competition placed greater weight on the organizational context and constraints in the Soviet Union, Marshall urged that we learn

> how the perceptions of what the other side is doing come about in various places within these complicated bureaucracies, and how these perceptions influence the behavior of the various organizations and the decision makers involved in the complex decision processes that drive . . . several defense programs.[16]

The research of Albert Wohlstetter on defense spending and arms programs revealed only partial connections between actions by one side and those by the other. He discovered that spending by Washington was not correlated to actions by Moscow. Moreover, he found that whereas the United States overestimated Soviet acquisition programs in some cases, in others it substantially underestimated them.[17] What is more, Ernest May, John Steinbruner, and Thomas Wolfe conducted an analysis of US-Soviet arms competition that discovered "budgets, forces, deployments, and policies of the United States . . . were products less of direct interaction with the Soviet Union than of the tension in the United States between dread of Communism on the one hand and the dread of deficit spending on the other."[18]

What is known about strategic interaction during the Cold War reveals that Soviet leaders paid attention to a mixture of external and internal developments. Their focus was selective and grasp of international affairs was mediated through strategic culture, ideology, and bureaucratic politics. Their responses were conditioned by organizational culture, bureaucratic politics, and standard operating procedures. Thus, the dynamics prevalent during the Cold War diverged significantly from expectations of the action-reaction theory.

Internal Sources of Competition

Arms competition is governed by domestic factors as well. A second set of theories deals with how national leadership, military establishments, and research-and-development bureaucracies affect decisions on arms procurement. The domestic structure model is based on the notion that forces within the state initiate arms competition. Such factors include the institutionalization of military research, development, and production; budgeting procedures; elections; the military-industrial complex; and varieties of organizational politics.[19]

First, national political objectives play a major role in arms decisions. In the case of China, for example, the need to coerce Taiwan while deterring, delaying, and defeating US intervention has been a major driver of Chinese arms investments. Bureaucratic politics also influences arms procurement. Memoirs of and interviews with military and industrial leaders in the Soviet Union reveal that competition between weapons design bureaus played a major role in determining the shape of investments in Soviet arms.[20] This is also likely to be the case with regard to China.

Military culture influences decisions on acquisition. Robert Perry, in the context of US-Soviet competition, writes that "whether Soviet or American, R&D institutions as readily aspire to organizational immortality as do trade guilds or cavalry regiments; instinctively, they resist change."[21] This is particularly the case in the United States, where the cultures of each military service influence what types of weapons they acquire.[22] Russia like the Soviet Union has separate branches for long-range missiles (Strategic Rocket Forces) and homeland air defense (Anti-Air Defense Troops). One favors intercontinental ballistic missiles (ICBMs) over submarine-launched ballistic missiles (SLBMs) and bombers, while the other heavily invests in air-defense interceptors and long-range surface-to-air missiles. The organizational culture of the People's Liberation Army (PLA) also influences acquisition. Specifically, the Second Artillery Force serves as a home and advocate for ballistic missiles within the Chinese military.

Bureaucratic processes also influence decisions on arms. For example, major dissimilarities in US and Soviet research, development, and acquisition systems produced significantly different forces.[23] The Soviets practiced evolutionary development, and the Americans preferred large-scale innovation.[24] Additionally, as Barry Buzan and Eric Herring write, "As the leading edge creates ever-higher standards of military capability, followers either have to upgrade the quality of their weapons or else decline in capability relative to those who do."[25] Solly Zuckerman observes, "The momentum of the arms race is undoubtedly fueled by the technicians in governmental laboratories and in the industries which produce the armaments."[26] Similarly, Marek Thee notes "the close interaction in stimulating the arms race between the feats of military technology, the interests and preferences of the military, the doings of the military industry and of the state political bureaucracy."[27] While the assertions may be inflated, research and development influence weapons decisions.

ARMS COMPETITION AND CHINA

Strategically, the United States and China are surely not autistic; they interact with one another. However, Sino-American strategic interaction is more complex than the action-reaction model would suggest. Even though Chinese leaders respond to US statements, programs, and actions, their attention is limited and selective. Moreover, it appears that domestic affairs, bureaucratic politics, organizational culture, and governmental practices exert influence.

Chinese decisions on weapons are obviously influenced by environmental considerations. Clearly the threat posed by the United States drove Chinese military modernization in the early years of the People's Republic of China. To counter that perceived nuclear threat, for example, Beijing required in 1965 that Dongfeng-3 (DF-3) intermediate-range ballistic missiles (IRBMs) be capable of reaching US bases in the Philippines and DF-4 IRBMs of striking Guam. China also called for the development and fielding of DF-5 ICBMs capable of hitting the continental United States.[28] In practice, the designers were not told to consider the strategic purpose of the weapons but instead were given the missile payload and range goals.

Chinese missile programs adapted only partially to the changed threat environment. Although ongoing programs experienced change after the Soviet Union was designated the primary enemy in the late 1960s, the basic features of the programs remained intact. Development of Soviet and American missile defenses also influenced the Chinese. As early as 1966, Qian Xuesen, head of the ballistic missile program, urged building penetration aids for DF-5 ICBM reentry vehicles with electronic countermeasures and light exoatmospheric decoys.[29]

After the end of the Cold War and collapse of the Soviet Union, the United States assumed a greater role in Chinese calculations. Several developments appear to have been particularly significant. The first event was the Gulf War in 1991. That conflict dramatically exhibited US prowess and called into question Soviet military doctrine and technology.[30] The performance of the US armed forces in that conflict and subsequent wars led Chinese analysts to conclude that the world was experiencing a revolution in military affairs stimulated by the advent of precision weaponry and information technology.

Chinese analysts thought that this development offered the opportunity to alter the military balance with the United States.[31]

Second, it appears that the Taiwan Strait Crisis of 1995–1996 served as an impetus for action by China. Specifically, inability of the PLA to detect US carrier battle groups, monitor operations near the Asian mainland, and deter or strike US power-projection forces spurred Chinese modernization. After this event, it became apparent that the capabilities of China to coerce Taiwan were inadequate. In particular, Chinese ballistic missiles could not ground the air force or knock out high-value targets on Taiwan. Moreover, Beijing lacked the wherewithal to hold American power projection at arm's length.[32] Much of the modernization by China in subsequent years was aimed at redressing this shortfall.

A third, underappreciated, incident was the accidental bombing of the Chinese Embassy in Belgrade in 1999 by US aircraft, which Beijing viewed as a deliberate act. According to General Zhang Wannian, vice chairman of the Central Military Commission, an emergency meeting in the wake of the bombing accelerated the development of *Shashoujian* (assassin's mace) weapons. Jiang Zemin, Communist Party general secretary, particularly argued in favor of such weapons because "what the enemy is most fearful of, this is what we should be developing."[33] Subsequently, China established the 995 Program to put defense and aerospace programs on a fast track and possibly include development of ASBMs.[34]

Although the strategic problem of coercing Taiwan while holding the United States at bay has driven Chinese modernization, other considerations also obtain. The assessment by Beijing of the security environment in Asia and its relative position to other powers including India and Russia likely play roles as well, as do less tangible but compelling matters such as the desire to appear as a great power equipped with state-of-the-art weapons systems.

Although external factors play important roles in Chinese decisions on arms, an exclusive focus on the security environment is insufficient to explain military modernization. Beyond the external environment, internal processes and dynamics influence the form and timing of its arms programs. The PLA culture as manifested in its lasting preferences influences the response.[35] For instance, developing antiaccess systems (*counterintervention* in the Chinese lexicon) such as precision-guided conventional ballistic missiles indicates one preference. Beijing has focused on ballistic missiles since the 1950s and created the Second Artillery as the organizational core of its nuclear missile capability. A Chinese defense white paper described the Second Artillery as "a strategic force under the direct command and control of the [Central Military Commission], and the core force of China for strategic deterrence."[36]

After discussion by the Central Military Commission, the missions of the Second Artillery were expanded in 1993 to add conventional missile deterrence and strike to nuclear deterrence and strike.[37] The missions are listed in doctrinal manuals such as *The Science of Second Artillery Campaigns.*[38] In fact, precision-guided conventional ballistic missiles have resulted in not only new missions but also the emergence of a group of officers who develop, deploy, and employ conventional strike systems. It is not clear, however, whether the nuclear and conventional officers possess distinct organizational cultures.[39]

Chinese leaders have also expressed an enduring preference for land-based mobile missiles. As Deng Xiaoping commented in August 1978, "I have the greatest interest in mobility on land; that is, in the use of modern weapons for fighting guerilla war."[40] The Julang 1 (JL-1) SLBM appears to have been dropped for the DF-21 mobile medium-range ballistic missile (MRBM) as land-based ballistic missiles were preferable to SLBMs.[41] Because of a propensity for concealment and deception, China heavily invested in a vast network of underground facilities—the Underground Great Wall—to protect weapons including missiles.[42]

Bureaucratic processes influence the scope and pace of arms acquisition, which include the assessment of the security environment codified in strategic guidelines; planning, programming, and budgeting; and research, development, and acquisition procedures. Military strategic guidelines issued by the Central Military Commission of the Central Committee of the Chinese Communist Party provide justification for modernization programs and establish priorities. They define the central military mission, identify potential enemies, and outline the scale and type of future conflicts. They have been revised only five times since the establishment of the People's Republic of China. They are issued in response to changes in the international or regional security environment, domestic affairs, and the character of war.

The current guidelines, "Military Strategic Guidelines for the New Period," were introduced by Jiang Zemin in 1993 as a result of an assessment of the international environment in the wake of the Cold War and demise of the Soviet Union and communist regimes in Eastern Europe as well as the changing nature of war after the Gulf War in 1991. The guidelines introduced *active defense*, which stressed precise and well-timed offensive operations, training, initiative, and attacks on enemy weaknesses. They were modified in 1999 after the Belgrade bombing, when the PLA began to characterize the most likely type of future war as "local wars under modern informationalized conditions."[43]

According to David Finkelstein, possibly "every modernization program, reform initiative, and significant change that the PLA has undergone . . . [is] the result of some of the fundamental decisions made when the new guidelines were promulgated in 1993."[44] Those programs include weapons developed and fielded under the Eighth (1991–1995), Ninth (1996–2000), Tenth (2001–2005), Eleventh (2006–2010), and Twelfth (2011–2015) Five-Year Plans.

Military strategy is developed on the basis of assessments of the security environment. In turn, the strategy helps in formulating defense requirements. The PLA General Armament Department makes acquisition decisions using a planning, programming, and budgeting system. During strategic research and development, it determines weapons requirements covering ten to twenty years that are instrumental in formulating the five-year armaments development plan with the details of research, development, and acquisition programs. Subsequently, the annual budget allocates the necessary resources to carry out the five-year plan.[45]

In the case of missiles, the development approach of the Chinese aerospace industry, which is characterized by incremental improvement, affects the structure and timing of research, development, and acquisition programs.[46] This strategy originated in the 1960s with a directive from the State Science and Technology director, Nie Rongzhen. It calls for having three variants of each weapons system in increasingly advanced stages of research and development at any one time: one in preliminary research; one in model research and development with design, development, testing, design reviews, and then finalization of the design; and one in low-rate initial production.[47] As new missiles go through this process, *seed units* become familiar with design teams, assembly plants, and supply chains. The units also develop tactics and maintenance procedures, as well as simulation systems, to ensure the smooth introduction of the new variant into the operational inventory.[48]

Bureaucracies also influence arms programs as military constituents pursue new missions. For example, the PLA Air Force and the Second Artillery desire to establish space operations as a core competency.[49] Furthermore, defense industries are seemingly involved in politicking. The First Academy of the China Aerospace Science and Technology Corporation (CASC) developed short-range ballistic missiles (SRBMs) in direct competition with the China Aerospace Science and Industry Corporation (CASIC), and there are indications that CASC is designing an MRBM to compete with systems developed by the CASIC Fourth Academy.[50]

Finally, it is likely that domestic affairs influence Chinese arms procurement. Although defense spending by Beijing has outpaced double-digit economic growth for many years, coffers in Beijing are not bottomless. Moreover, defense has competed against other budgetary priorities. Like other categories of national spending, it provides money and jobs that benefit the domestic economy. As Zhang Yunzhuang of the Chinese National University of Defense Technology writes, procurement is affected by bureaucratic and domestic politics, including the pressure to buy older weapons to help state-owned enterprises avoid bankruptcy.[51] Such considerations likely affect modernization more profoundly than generally appreciated.

ACTION-REACTION OR STRATEGIC AUTISM?

The more closely arms programs are examined, the more difficult it becomes to discover evidence of a tightly coupled action-reaction arms race. Although Beijing is not strategically autistic, internal structures and processes are critical to making decisions on weaponry. This dynamic is illustrated by the advent of precision-guided conventional missiles, including ASBMs.

On one level, Chinese conventionally armed precision-guided ballistic missiles seem to be a response to US power projection using forward air bases and carrier strike groups. However, when further examined the situation is more complex. The weapons being fielded now are the result of decisions taken years or decades ago under quite different external and internal conditions. The precision-guided conventional ballistic missiles, for instance, were not originally intended to threaten bases in Taiwan or the United States. DF-15 SRBMs developed by CASC and DF-11 SRBMs developed by CASIC date from the 1980s and began as an effort to counter Soviet deep operations and increase arms exports.

China developed conventional ballistic missiles to counter threats of a deep combined-arms attack by the Soviet Union on its northern border. Missiles were an attractive way to supplement relatively weak capabilities to mount air strikes and rapidly target enemy rear areas. The Chinese space and missile industry in late 1984 shifted its emphasis from missiles with liquid to solid fuel and from weapons with strategic to tactical range.[52]

Another rationale for developing conventional ballistic missiles was promoting arms sales to the developing world. Research and development on DF-15s and DF-11s began in 1985. Active international marketing of DF-15 missiles began in 1986, and Syria agreed to purchase them in 1988. In addition, DF-11 missiles were successfully flight tested in 1990 and a contract for their sale to Pakistan was signed the following year.[53]

In 1988 the PLA decided to deploy conventional missiles opposite Taiwan and in 1991 organized a seed unit to accept them.[54] Base 52, the main missile base adjacent to Taiwan, accepted the first DF-15s in 1992. A conventional missile brigade was commissioned the next year and was ordered to be operational within a year.[55] Between the mid-1990s and 2010, the conventional SRBM arsenal was enlarged from 30–50 inaccurate missiles to 1,000–1,200 more accurate and lethal SRBMs. They included 350–400 DF-15 missiles and 700–750 DF-11 SRBMs.[56] Turning to the development of ballistic missiles, DF-21s and their variants, the link between the influence of the external environment and decisions on military procurement has become more apparent.

The development of DF-21D ASBMs would seem to illustrate clearly an action-reaction dynamic. Certainly the available evidence shows that DF-21D development was driven by the requirement to hold US carrier strike groups at bay to deter future intervention. The near-term goal called for keeping carriers at least two thousand kilometers from the coast, beyond the range of carrier-based strike aircraft and *Tomahawk* land-attack cruise missiles. As one analyst notes, Beijing is "obsessed with defending China from long-range precision air strikes."[57] Moreover, pertinent research appears to have begun in 1996 after the Taiwan Strait Crisis. That year CASC started ASBM feasibility studies and concept demonstrations, particularly in guidance and control systems.[58]

However, the shape and timing of the program also have been driven by internal dynamics. The DF-21D is only the most recent variant of the DF-21 missile program, which began with the fielding of the JL-1 SLBM to counter the Soviet threat. Nevertheless, the JL-1 program revealed deficiencies in defense industries, including missile components such as solid rocket motors, structures, warheads, and reentry vehicles. The lack of pressing strategic issues to drive innovation hampered the program. Specifically, the Chinese remained uncertain about ballistic-missile submarines (SSBNs) armed with SLBMs because of strategic geography: they doubted submarines could get within firing range of the Soviet Union and were concerned that the waters around China were too shallow to conceal them.

As a result of Chinese strategic geography and the PLA organizational culture, in the early 1970s the leadership hedged its bets by developing the land-based version of the JL-1, called the DF-21 (CSS-5 Mod-1) MRBM. The two systems were to be developed in tandem with the same airframes and engines. DF-21s were eventually favored over the sea-based version and entered the operational inventory of the Second Artillery in 1991 with a regional nuclear strike mission, gradually replacing the older liquid-fueled DF-3A IRBMs.

In keeping with the incremental development process in the aerospace sector, China began the DF-21A (CSS-5 Mod-2) MRBM program in 1987 to lower structural weight, add propellant, and boost the second-stage thrust to increase the range by 60 percent. The design work started in 1988 and the initial tests in 1992. DF-21As tested to a three-thousand-kilometer range, and the design was finalized in 1997, with deployment in the late 1990s and 2000s.

The DF-21C (CSS-5 Mod-3) precision-guided conventional ballistic missile was the next system deployed. China became interested in precision-strike technology in the late 1970s and studied the *Pershing II*, publishing more than fifty articles on the missile technology. In addition, precision weaponry benefited from the 863 Program, which was established in 1986 to improve nuclear deterrence in the face of US missile defenses. China also formed the Precision Guidance Expert Group to focus on terminal guidance, leading to a modernization program spanning more than four decades that produced DF-21D ASBMs.[59] The influence of internal structures and processes is so marked in the case of the DF-21D that it should give pause to anyone who characterizes Sino-American competition as a single-minded action-reaction arms race.

• • •

Neither the United States nor China is strategically autistic; they pay attention to, and respond to, the external environment. Moreover, they compete against one another. China has competed with the United States for two decades, while the United States has paid increased attention to China in more recent years. However, their competition falls short of the type of action-reaction arms race described by international relations theorists during the Cold War. Even though the United States and China are building force structures with reference to one another, competition between them is not the sole or main driver of their respective modernization efforts.

Although an arms race between the United States and China or between China and some other regional actor might emerge in the future, it is hardly preordained. Even if one does occur, internal factors such as organizational culture, institutional processes, bureaucratic politics, and domestic issues are likely to occupy prominent roles in determining the scope and pace of the competition. American scholars and policy makers must develop wide-ranging knowledge of Chinese history, culture, politics, and organization to fully grasp this important dynamic.

NOTES

1. Desmond Ball, "Asia's Naval Arms Race: Myth or Reality?" (paper presented at the twenty-fifth Asia-Pacific Roundtable, Shangri-la Hotel, Kuala Lumpur, Malaysia, May 30, 2011). See also Geoffrey Till, *Asia's Naval Expansion: An Arms Race in the Making?* (London: Routledge for the International Institute for Strategic Studies, 2012); and Brad Roberts, *Asia's Major Powers and the Emerging Challenges to Nuclear Stability Among Them*, IDA Paper P-4423 (Alexandria, VA: Institute for Defense Analyses, 2009).

2. Edward N. Luttwak, *The Rise of China vs. the Logic of Strategy* (Cambridge, MA: Belknap Press, 2012).

3. Samuel P. Huntington, "Arms Races: Prerequisites and Results," *Public Policy* 8 (1958): 41.

4. Colin S. Gray, "The Arms Race Phenomenon," *World Politics* 24 (October 1971): 40.

5. Barry Buzan, *An Introduction to Strategic Studies: Military Technology and International Relations* (New York: St. Martin's, 1987), 73.

6. Barry Buzan and Eric Herring, *The Arms Dynamic in World Politics* (London: Lynne Rienner, 1998), 83.

7. George W. Rathjens, *The Future of the Strategic Arms Race: Options for the 1970s* (Washington, DC: Carnegie Endowment for International Peace, 1969), 25–26.

8. Robert Jervis, "Cooperation Under the Security Dilemma," *World Politics* 30 (January 1978): 169.

9. Charles L. Glaser, "The Security Dilemma Revisited," *World Politics* 50 (October 1997): 192.

10. Paul M. Kennedy, *The Rise of the Anglo-German Antagonism, 1860–1914* (London: Allen and Unwin, 1980).

11. Charles L. Glaser, "The Causes and Consequences of Arms Races," *Annual Review of Political Science* 3 (June 2000): 254.

12. Buzan and Herring, *The Arms Dynamic*, 83–86.

13. Colin S. Gray, "The Urge to Compete: Rationales for Arms Racing," *World Politics* 26 (January 1974): 210–27.

14. Quoted in Glaser, "The Causes and Consequences of Arms Races," 253.

15. Rathjens, *The Future of the Strategic Arms Race.*

16. Andrew W. Marshall, *Long-Term Competition with the Soviets: A Framework for Strategic Analysis*, R-862-PR (Santa Monica, CA: RAND, 1972), 7.

17. Albert Wohlstetter, "Is There a Strategic Arms Race?" *Foreign Policy* 15 (Summer 1974): 3–20; Albert Wohlstetter, "Rivals, but No 'Race,'" *Foreign Policy* 16 (Fall 1974): 48–81.

18. Ernest R. May, John D. Steinbruner, and Thomas W. Wolfe, *History of the Strategic Arms Competition, 1945–1972*, part 1 (Washington, DC: Historical Office, Office of the Secretary of Defense, 1981), 241.

19. Buzan and Herring, *The Arms Dynamic*, chap. 7.

20. David Holloway, "The Soviet Style of Military R&D" in *The Genesis of New Weapons: Decision Making for Military R&D*, ed. Franklin A. Long and Judith Reppy (New York: Pergamon Press, 1980), 145.

21. Robert Perry, "American Styles of Military R&D," in Long and Reppy, *Genesis of New Weapons*, 107.

22. Thomas G. Mahnken, *Technology and the American Way of War Since 1945* (New York: Columbia University Press, 2008).

23. Franklin A. Long and Judith Reppy, "Decision-Making in Military R&D: An Introductory Overview," in Long and Reppy, *Genesis of New Weapons*, 13–14.

24. Holloway, "The Soviet Style of Military R&D," 139.

25. Buzan and Herring, *The Arms Dynamic*, 30.

26. Solly Zuckerman, *Nuclear Illusion and Reality* (London: Collins, 1982), 103.

27. Marek Thee, "Military Technology: A Driving Force Behind the Arms Race and an Impediment to Arms Control and Disarmament," in *Military Technology, Armaments Dynamics and Disarmament*, ed. Hans Günter Brauch (New York: St. Martin's Press, 1989), 42.

28. John Wilson Lewis and Hua Di, "China's Ballistic Missile Programs: Technologies, Strategies, Goals," *International Security* 17 (Fall 1992): 14–17.

29. Ibid., 20.

30. Andrew Scobell, David Lai, and Roy Kamphausen, eds., *Chinese Lessons from Other Peoples' Wars* (Carlisle, PA: Strategic Studies Institute, US Army War College, 2011).

31. Jacqueline Newmyer, "The Revolution in Military Affairs with Chinese Characteristics," *Journal of Strategic Studies* 33 (August 2010): 483–504.

32. David Shambaugh, "China's Military Modernization: Making Steady and Surprising Progress," in *Strategic Asia, 2005–06: Military Modernization in an Era of Uncertainty*, ed. Ashley J. Tellis and Michael Wills (Seattle: National Bureau of Asian Research, 2005).

33. *Zhang Wannian Chuan* [Biography of Zhang Wannian] (Beijing: Liberation Army Press, 2011), 416, 419. I am grateful to Professor Tai Ming Cheung of the University of California, San Diego, for bringing this to my attention.

34. Andrew S. Erickson, *Chinese Anti-Ship Ballistic Missile (ASBM) Development: Drivers, Trajectories and Strategic Implications* (Washington, DC: Jamestown Foundation, 2013), 21; Mark A. Stokes, *China's Evolving Conventional Strategic Strike Capability* (Washington, DC: Project 2049 Institute, 2009), 27.

35. Definitions of *strategic culture* abound. Andrew Scobell considers it "values held in common by the leaders or group of leaders of a state concerning the use of military force." "Soldiers, Statesmen, Strategic Culture, and China's 1950 Intervention in Korea," *Journal of Contemporary China* 8 (1999): 479. The Defense Threat Reduction Agency suggests that it is "shared beliefs, assumptions, and modes of behavior, derived from common experiences and accepted narratives . . . which determine appropriate ends and means for achieving security objectives." Jeffrey S. Lantis, "Strategic Culture: From Clausewitz to Constructivism," in *Strategic Culture and Weapons of Mass Destruction: Culturally Based Insights into Comparative National Security Policymaking*, ed. Jeannie L. Johnson, Kerry M. Kartchner, and Jeffrey A. Larsen (New York: Palgrave Macmillan, 2009), 39. Alastair Iain Johnston defines *strategic culture* as a "system of symbols (e.g., argumentation structures, analogies, metaphors) which acts to establish pervasive and long-lasting strategic preferences by formulating concepts of the role and efficacy of military force in interstate political affairs." "Thinking About Strategic Culture," *International Security* 19 (Spring 1995): 46.

36. People's Republic of China, *China's National Defense in 2008* (Beijing: Information Office of the State Council, 2009), 40.

37. John W. Lewis and Xue Litai, "Making China's Nuclear War Plan," *Bulletin of the Atomic Scientists* 68 (2012): 52.

38. Second Artillery Corps of the People's Liberation Army, *The Science of Second Artillery Campaigns* (Beijing: People's Liberation Army Press, 2004).

39. Ron Christman, "Conventional Missions for China's Second Artillery Corps," *Comparative Strategy* 30 (July–August 2011): 213.

40. Lewis and Hua, "China's Ballistic Missile Programs," 26.

41. Ibid., 21.

42. Phillip A. Karber, *Strategic Implications of China's Underground Great Wall* (Washington, DC: Asian Arms Control Project, Georgetown University, 2011).

43. David M. Finkelstein, "China's National Military Strategy: An Overview of the 'Military Strategic Guidelines,'" *Asia Policy*, no. 4 (July 2007): 69.

44. Ibid.

45. Zhang Yunzhuang, "China's Military Procurement and Its Operational Implications: A Response to Yoram Evron," *Journal of Strategic Studies* 35 (December 2012): 891–92.

46. Mark A. Stokes and Ian Easton, *Evolving Aerospace Trends in the Asia-Pacific Region: Implications for Stability in the Taiwan Strait and Beyond* (Washington, DC: Project 2049 Institute, 2010), 9.

47. Stokes, *China's Evolving Conventional Strategic Strike Capability*, 10.

48. Stokes and Easton, *Evolving Aerospace Trends*, 9.

49. Mark A. Stokes and Dean Cheng, *China's Evolving Space Capabilities: Implications for US Interests* (Washington, DC: Project 2049 Institute, 2012), 45.

50. Ibid., 17–18.

51. Zhang, "China's Military Procurement," 892.

52. Christman, "Conventional Missions," 202.

53. Stokes and Easton, *Evolving Aerospace Trends*, 10; Lewis and Hua, "China's Ballistic Missile Programs," 34–36.

54. Stokes and Easton, *Evolving Aerospace Trends*, 13.

55. Lewis and Xue, "Making China's Nuclear War Plan," 53.

56. Michael S. Chase and Andrew S. Erickson, "A Competitive Strategy with Chinese Characteristics? The Second Artillery's Growing Conventional Forces and Missions," in *Competitive Strategies for the 21st Century: Theory, History, and Practice*, ed. Thomas G. Mahnken (Palo Alto, CA: Stanford University Press, 2012), 209.

57. Quoted in Stokes, *China's Evolving Conventional Strategic Strike Capability*, 6.

58. Erickson, *Chinese Anti-Ship Ballistic Missile (ASBM) Development*, 410–45.

59. Ibid., 8.

14 IRREGULAR WARFARE IN ASIA

Michael Evans

A STUDY OF IRREGULAR warfare in the context of Asian strategic studies leads to three conclusions. First, modern strategic studies have addressed irregular warfare in ways that suffer from undertheorizing and lack of historical perspective. Second, the interest in irregular warfare in Asia is focused on South and Southeast Asia. Third, while research on Asian irregular warfare must be sensitive to the realities of strategic culture, exotica like Asian values should be avoided. Much can be learned from the character of diverse irregular conflicts that have plagued some parts of Asia, but in this century most analytical techniques associated with counterinsurgency (COIN) and counterterrorism (CT) are grounded in Western models of strategic studies.

However, irregular warfare has often been neglected in modern strategic studies.[1] Part of this pattern of neglect can be attributed to the difficulty in finding a workable definition. Also, the diverse nature of irregular warfare defies systematic theorizing, reducing much strategic analysis to the operational and tactical aspects of sui generis regional or national issues. Finally, subcomponents of irregular warfare are analyzed in their own right. As one authority frames this issue, "Irregular warfare is the umbrella term used to describe violence used by sub-state actors . . . including terrorism and insurgency."[2]

MODERN STRATEGIC STUDIES

Insurgency and terrorism are related and employed independently or jointly to overthrow or paralyze the established order. Even though these subcomponents represent specific types of unconventional warfare and low-intensity conflict, their course from Algeria to Ulster offer some degree of theoretical stability for analysis. Thus, it can be deduced that irregular warfare in the twentieth century blended insurgency and terrorism and has mutated further under the influence of globalization and electronic networks in the new millennium.[3]

Since the emergence of modern strategic studies in the mid-1950s, most intellectual efforts have been focused on developing methods of COIN and CT, which are necessary to defeat violent substate groups that range from Marxist revolutionaries and national liberation movements to separatist elements. Two separate COIN eras stimulated unconventional warfare analysis in Western strategic studies. The first one occurred from the early 1950s to the mid-1970s and was shaped by three major events: people's war in China, insurgencies in the Third World that employed Maoist revolutionary warfare, and ideological struggles of the Cold War and the accompanying nuclear stalemate that favored the use of COIN as an antidote to the spread of people's war.[4]

By the end of the 1970s, this first COIN era had waned along with European colonialism. The outcome favored the Third World as Cold War rivalries pursued détente, arms control, and new precision weapons that weakened the nuclear stalemate. With the exception of the 1948–1960 Malayan campaign, that great exemplar of COIN operations, Western experience with irregular warfare usually resulted in military frustration and political defeat. Indeed, a number of scholars who studied the first COIN era pilloried Western efforts: "Nothing in contemporary military writing is more pathetic than the voluminous attempts to find viable counterinsurgency doctrines."[5] Lamenting this state of affairs, they suggested "The syncretizing of archaic and modern forces in insurgency has never been properly grasped in the considerable literature on the subject—most of it dealing with counterinsurgency practices."[6]

Such views proved influential in the marginalization of irregular conflict into the realm of what one observer calls "uncomfortable wars to be avoided at all costs."[7] Not surprisingly, both Western strategic studies and defense policy makers were ill prepared for the 9/11 attacks and the US-led campaigns in Iraq and Afghanistan. This lack of awareness and readiness during the last decade has been dramatically highlighted by a group of academics who were moved to observe, "The major intelligence failure . . . was the startling lack of attention given to the rise of irregular warfare—including insurgency, warlordism and the 'new terrorism.'"[8]

Given the need for remedial action, irregular conflict after 2004 came to resemble a catch-up game in which a deficit score is fiercely pursued. As a result, academic studies and doctrinal development on the subject went from bust to boom. Over recent years an avalanche of Western studies representing a second COIN era has filled the vacuum in knowledge. In fact, "a conceptual reorientation of gigantic proportions" has occurred because of Afghanistan and Iraq, which has rejuvenated COIN theories to "almost encyclopaedic proportions."[9] Classical writings by such practitioners as Robert Thompson, Frank Kitson, and David Galula have been rediscovered for their wisdom on conducting population-centric COIN operations. Different religious, cultural, anthropological, and historical insights also have been injected into strategic studies. And a new generation of theorists has surveyed irregular warfare for a postindustrial world in which insurgents roam from London to Bombay.

The conceptual blending of terrorism and insurgency has become central to understanding irregular warfare. The US Department of Defense *Joint Operating Concept* states that all activities must be "undertaken in sequence, in parallel, or in *blended form* in a coherent campaign to address irregular threats."[10] Moreover, the Department of Defense defines *irregular warfare* as "a violent struggle among state and non-state actors for legitimacy and influence over the relevant populations. [It] favors indirect and asymmetric approaches, though it may employ the full range of military and other capabilities, in order to erode an adversary's power, influence and will."[11]

The Pentagon has acknowledged that irregular warfare is just as important strategically as traditional warfare and construed it to include CT, unconventional warfare, foreign internal defense, COIN, and stability operations against state and nonstate actors across a spectrum of conflict.[12] In the foreseeable future, regardless of the outcome in Afghanistan, it is unlikely that irregular warfare will be relegated again to the margins of strategic studies or doctrinal backwaters in Western nations.

SOUTH AND SOUTHEAST ASIA

With the connection between irregular warfare and strategic studies established, it is possible to examine two areas in a regional context with emphasis on South and Southeast Asia rather than East Asia. While Mao Zedong produced one of the seminal texts of the last century on irregular warfare, today Beijing is bent on high-tech military modernization aside from its dalliance with unrestricted warfare. As a result, much contemporary scholarship has focused on the old security agenda of high-tech conventional warfare, military modernization, and nuclear weaponry surrounding the rise of China in East Asia.[13] Comparatively less attention has been paid to the new security agendas of South and Southeast Asian irregular warfare and its subcomponents of insurgency and terrorism, which embrace a volatile mixture of Islamist extremism, transnational threats, and weak states in South Asia and parts of Southeast Asia.

Asia may become a future laboratory for crosscutting themes on the old security agenda of modern geopolitics and interstate rivalry and the new security agenda of postmodern globalized security and nonstate threats. For example, major nations such as China, India, Pakistan, and the United States have important interests in irregular conflicts in South and Central Asia that range from Afghanistan and Kashmir to Xingjiang. Moreover, insurgencies throughout Southeast Asia and various radical Islamist movements threaten the stability of several vital nations, including Indonesia, which is the great Muslim pivot nation of the maritime subregion.[14]

This situation must be considered against this background of twenty-first-century security agendas found in those nations most challenged by irregular adversaries—namely, India, Pakistan, and Sri Lanka. Faced by diverse insurgencies and terrorist campaigns, each of these nations has tailored its response to national requirements, from India's population-centric COIN to the more enemy-centric strategies of Pakistan and Sri Lanka.

India

An enormous ethnic diversity has been held in check within India by a strong democratic culture and the tolerant spirit of Nehruvian federalism. For over sixty years, Indian democracy has faced serious challenges from secessionist insurgencies and sectarian terrorism, but despite such crises it has endured. During the second half of the last century these challenges included insurgencies in Nagaland and Mizoram in northeast India, Sikh separatism in the Punjab, and intervention in Sri Lanka from 1987 to 1990.[15] The Sikh demand for a separate state called Khalistan led to a struggle that claimed more than twenty thousand lives, which exceeded the deaths in Indo-Pakistan wars and included a prime minister and a former chief of staff. As C. Christine Fair notes, the Sikh insurgency stands out because it was completely defeated.[16]

In the new millennium, the Naxalite Maoist threat and the rise of proxy warfare in Kashmir have emerged as serious challenges to the Indian state. The Naxalite communists who originated in West Bengal and Adrah Pradesh have been particularly dangerous in recent years. Since 2004, following the unification of several factions into the Maoist Communist Party of India, Naxalites have emerged, according to Prime Minister Manmohan Singh, as the nation's greatest internal threat. Today, this insurgency exploits poverty and landlessness in eighteen states. In addition, unlike other conflicts, this pan-Indian movement promotes socialist revolution.[17]

Most irregular challenges to India since independence have been domestic and entrenched rather than foreign and recent in origin. Sri Lanka and Kashmir involve major external problems with either expeditionary operations in the former or confronting proxy forces and *jihadis* in the latter. The intervention in Sri Lanka in 1987 represented the only occasion when India embarked on an overseas counterinsurgency campaign.[18] While these operations were mounted initially to enforce a peace accord between the Sinhalese government and Tamil Tigers, Indian troops soon confronted Tamil fighters on the Jaffna Peninsula. As one observer notes, through both regular and irregular tactics the terrorists "roughed up the fourth largest army in the world." The Indian Army was found wanting in COIN warfare, including politicomilitary coordination, command and control, logistics, intelligence, and antimine warfare.[19]

India confronted the *new* in irregular warfare in Kashmir with the global *jihadi* movement. While that insurgency originated as an indigenous phenomenon linked to Indo-Pakistan rivalry, it morphed from a local insurgency into a proxy war. Pakistan used Kashmiri Islamists such as Hizb-ul-Mujahideen who inclined toward a union with Pakistan to weaken the separatist Jammu Kashmir Liberation Front.[20] India's containment of the *Kashmiriyat* (the distinct cultural values of Kashmir) insurgency led to the Pakistanization of the Kashmir crisis in the 1990s. According to one observer, "Rather than being an insurgency per se, [Kashmir] became connected with . . . an unconventional warfare campaign, the use of Pakistani intelligence (and special operations) personnel to train external jihadis to augment indigenous rebels."[21]

At the same time, al-Qaeda operatives also extended assistance to Islamist Kashmiri groups such as Harkat-ul-Mujahideen, Harkat-ul-Jihad, Jaish-e-Mohammed, and Lashkar-e-Toiba. What started in the 1980s as a classical anti-Indian insurgency in pursuit of *azaddi* (freedom) became a proxy war by 2004 in which jihadists made up more than 80 percent of the casualties. Groups with global connections such as Lashkar-e-Toiba attacked the legislatures of Kashmir and India in 2001 and then executed the deadly raid in Mumbai in 2008.[22]

Unlike other insurgencies in India, the security situation in Kashmir underwent a double transformation, from a separatist insurgency to a Pakistani proxy war and then to a form of globalized *jihadi* irregular conflict. Currently, the support for the insurgency in Kashmir comes from not the local population but rather foreign actors. In general, the Indian response to irregular threats has been influenced by Western methods that honor liberal democratic values.[23] As Sumit Ganguly and David Fidler find, no Indian way of war exists in irregular operations, and COIN reinforces traditional warfighting.[24] Moreover, the influence of British experts like Robert Thompson and Frank Kitson is strong. For example, it became an article of faith in the Indian Army that the object of COIN is managing violence "to destroy the political organization that sustains a guerrilla movement."[25]

Pragmatism also fostered ad hoc methods and reliance on experience over doctrine that led to costly relearning in many contingencies. Indeed, curious blends of the velvet glove, iron fist, and military amnesia have emerged, which produced a Zen-like paradox: "The more India wages COIN, the less it really learns."[26] British approaches that derived from the Malayan Emergency (1948–1960) and emphasized policing and intelligence were transferred to India, although other methods did not. For example, population resettlement in Nagaland proved to be counterproductive, unlike in Malaya, and was dropped from the COIN repertoire.[27]

In developing COIN doctrine, India faced the dilemma of other democracies, which is achieving a balance between the use of military force and upholding the rule of law and human rights. Unsurprisingly, an analysis of Indian COIN reveals that this tension contributed to methodological paradoxes. According to one commentator, India's sprawling federal democracy often makes civil-military doctrine dysfunctional, with "the only conceptual Indian government template for understanding and dealing with internal rebellions [being] Nehru's thinking on these issues in the 1950s."[28]

Despite a fondness for political solutions, the provisions of the Armed Forces Special Powers Act of 1958, which indemnifies soldiers from criminal prosecution, have long bolstered Indian COIN. This legislation has been criticized by the media and civil libertarians as antithetical to the rule of law and human rights.[29] COIN has resulted in heavy-handed measures, notably storming of the Sikh Golden Temple in Amritsar in 1984 with the loss of more than four thousand lives. However, most Indian COIN efforts have been civilized and tempered, using a law enforcement "strategy of political flexibility, negotiation, and principled compromise."[30]

In 2006 India codified its long experience in CT, insurgency, and proxy wars in *Doctrine for Sub-Conventional Operations* in 2006. This canon sought to capture the collective wisdom acquired over almost five decades in fighting subconventional warfare.[31] The doctrine reveals the continuing influence of Thompson and Kitson on COIN in India by emphasizing the primacy of a civil face, use of minimum force, and upholding the view of the people as the center of gravity in all subconventional operations. Unity of command and civil-military relations that have been weak in the past are upheld by the Apex Body to supervise all agencies in a conflict.[32] Ultimately, the most distinguishing feature of COIN is the Indian tradition of political flexibility and eliminating insurgencies, but not the insurgents.[33] As a huge democracy seeking to maintain the vision of pan-Indian unity, the nation has favored negotiated solutions to grievances that fuel insurgencies. Because COIN facilitates the political integrity of India, it should be acknowledged as a major success despite paradoxes and contradictions in applying its lessons to domestic security challenges.

Pakistan

If India has succeeded in developing a complex multireligious democratic nation, Pakistan has demonstrated exactly the opposite. The secular state established by Mohammed Jinnah in 1947 has over several decades descended into frequent military rule even while becoming a nuclear weapons state and facing the specter of Islamist extremism. An unhappy combination of military rule, strategic rivalry with India, instability in its neighbor Afghanistan, and Islamist influences has indelibly shaped Pakistan's view of irregular warfare. In the twenty-first century, the nation has a schizophrenic strategy that has led it to become a sanctuary for jihadist proxy war against India and a de facto ally of the West against the very movement it tolerates on its soil.[34]

Some elements in the Pakistani Directorate for Inter-Services Intelligence (ISI) tolerate and support irregulars and terrorists, which resulted in the US special operations raid into Pakistan to eliminate Osama bin Laden in 2011. The syndicate that is present in Pakistan includes senior al-Qaeda leaders, the Pakistani Taliban and Haqqani networks, the Afghan Taliban conducting the war against the US-led International Security Assistance Force (ISAF), and the Lashkar-e-Toiba, which is focused on creating a Mughal caliphate. In addition, groups such as Lashkar-e-Janghvi, Sipa-e-Sohaba Pakistan, and Jaish-e-Mohammad are supported by fighters recruited from countries ranging from Morocco and Yemen to Chechnya and Indonesia.[35] As Bruce Riedel comments, in the twenty-first century Pakistan represents "the birthplace of global Islamic jihad and [is] now its epicentre. [It is] a crucible of terror."[36]

Strategists of global jihad from around the Muslim world, ranging from the Saudi Osama bin Laden and Abd al-Aziz al-Muqrin, the Palestinian Abdallah Yisuf Mustafa Azzam, and the Syrian *jihadi* Abu Mus'ab al-Suri, have found refuge in Pakistan.[37] President Barack Obama accurately described the Pakistani border region as the most dangerous place in the world in terms of Western interests.[38] It is on the Pakistan-Afghan frontier that the diverse Pakistani and foreign terrorist, insurgent, and criminal groups have established "an archipelago of micro emirates imposing *sharia* across large swaths of the Pashtun belt."[39]

Given the blatant ISI record of harboring *jihadis*, the post-2009 COIN efforts of the Pakistani Army in border regions against Islamist militants can only be seen as a kind of hedging strategy.[40] This struggle in the northwest, especially the Swat Valley, has seen more Pakistani soldiers killed fighting Taliban irregulars than ISAF troops. Echoing President Obama, Bruce Riedel characterizes the festering threat: by 2010 "the most dangerous place in the world for America was the border badlands between Pakistan and Afghanistan."[41]

A major factor that affects the ISAF mission in Afghanistan is the Pakistani preoccupation with India and concern over long-term strategic interests, which combined have led to a neglect of COIN in Waziristan and the Swat Valley. Pakistani COIN often has been enemy-centric and focused on creating free-fire zones that have displaced four hundred thousand civilians and generated widespread resentment. As one observer notes, "The Pakistan Army's COIN strategy has often verged on the myopic. . . . It has still not sunk in that the Taliban are a more immediate existential threat to Pakistan's security than India."[42]

Pakistani failures have enabled the cross-border movement of insurgents, which has been mitigated by ISAF drone attacks to kill militant leaders of the Pakistan Taliban and al-Qaeda in Afghanistan.[43] It is difficult not to agree with Riedel that "a jihadist Pakistan would be the most serious threat . . .

since the end of the Cold War. Aligned with al-Qaeda and armed with nuclear weapons, the Islamic Emirate of Pakistan would be a nightmare."[44]

Sri Lanka

Of all separatist South Asian insurgent movements, arguably that of the Tamil Tigers, who fought the Sinhalese majority in Sri Lanka, was the most ruthless and formidable. They have been highly adaptive and creative terrorists, whose tactics have been adopted by insurgent organizations around the world. Led by Vilupillai Prabhakaran from 1983 to 2009, the Tamil Tigers developed improvised explosive devices and suicide bombs, killing more people than Hamas and Hezbollah combined, including Rajiv Gandhi in 1991.[45]

Although the Tamil Tigers were a separatist nationalist movement, members of the Tamil diaspora provided money and weapons through what by the beginning of the twenty-first century had become a global network to sustain the insurgents in their stronghold on the Jaffna Peninsula. Yet for all their deadly skill, the Tamil Tigers represent one of the few groups to be eradicated using military forces in the application of an enemy-centric approach to COIN. The reason for the demise of this group can be found in its very success. As an ethnic separatist movement with ambitions for statehood, it became a hybrid warfare movement fielding not only irregular forces but also militias and regulars able to fight positional battles.[46]

When President Mahinda Rajapaska and Army Chief Lieutenant General Sarath Fonseka choose a strategy of eradication in 2005, ethnic civil war rather than COIN ensued. The extent of the fighting can be appreciated by the fact that the Sri Lankan Army deployed fifty-nine divisions.[47] But its victory in 2009 was achieved in ways that could not be sanctioned by liberal democracies: media blackouts, massive firepower, and indifference to civilian deaths. As one Indian writer notes, "Sri Lanka has set a new paradigm on the use of force, but incurred huge humanitarian and diplomatic costs for its all-out use of force."[48]

Southeast Asia

If South Asia has been seen as a first front for modern irregular challenges, Southeast Asia has been characterized as a second front. This is unsurprising given that volunteers from Indonesia, Malaysia, the Philippines, and Thailand were trained by al-Qaeda in Afghanistan in the 1990s. Mass casualty attacks on Bali and in Jakarta in the new millennium provided graphic evidence that the extremist jihadist influence exists throughout Southeast Asia.[49]

However, like most of the Indian subcontinent, Southeast Asia has its own historical dynamic that reflects long experience with secessionist and separatist movements. Given the mix of sultanates, local polities, and ethnic groups, the political evolution of Southeast Asia has been marked by internal conflicts against centralization, ranging from the Philippines to Thailand and from Indonesia to East Timor.[50] The issues of legitimacy in nation building, authoritarian rule, and economic underdevelopment played roles in undermining unity in multiethnic states of Southeast Asia. According to Muthiah Alagappa, the contested political legitimacy of many regimes, from Ngo Dinh Diem in South Vietnam through assorted generals in Thailand to the regime of the Indonesian General Suharto, has fashioned a systemic problem with internal security across the region. Consequently, guerrillas proliferated, from the Khmer Rouge in Cambodia, Darul Islam in Indonesia, the Moro National Liberation Front and Abu Sayyaf in the Philippines, to armed groups in Burma, Laos, and Thailand.[51]

Another similarity with South Asia has been the coexistence of the old (traditional ethnonationalist insurgencies) and the new (transnational *jihadi* Islamism) in irregular challenges throughout Southeast Asia. In the early twenty-first century, older separatist insurgencies such as the Muslim Moros in the Philippines, the Pattani Malays in Thailand, and the Acehnese in Indonesia coexist with newer millenarian and transnational jihadi movements such as Jemaah Islamiyah and Laskar Jihad. Like Indian Naxalites in West Bengal today, the Maoist New People's Army was capable of mobilizing ten thousand guerrillas in Luzon in 2004.[52]

Irregular conflict in Southeast Asia is complex in that it "cannot be viewed in narrow definitional terms nor is it amenable to a set of generalizations, and hence narrow prescriptive countermeasures."[53] A blend of local and global dynamics is vital to understanding the region. For example, although Jemaah Islamiyah originated in Indonesia during the 1950s in the local Darul Islam rebellion, it has formed a twenty-first-century association with al-Qaeda and other global *jihadi* groups. Similarly, Moro fighters in Mindanao in the Philippines also have external links, with some serving as volunteers in Afghanistan, and in 2004 the al-Qaeda-linked Abu Sayyaf group blew up a Philippine ferry killing more than a hundred people. *Jihadi* inroads in Southeast Asia have not, however, always been successful. For example, the Moro Islamic Liberation Front severed its links with external *jihadis* to avoid diluting its agenda.[54]

In Thailand's southern provinces, Pattani Muslim separatism is strong, with groups such as the Gerakan Mujahideen Islam Pattani having linkages to al-Qaeda and other groups. However, most Pattani separatists have avoided al-Qaeda and Lashkar-e-Toiba-style mass-casualty attacks, realizing that such an approach inflames the Buddhist majority.[55] However, the COIN campaign has been intense and bloody at times. For example, in April 2004, the Thai military killed 108 Muslim guerrillas in a single day. As Andrew Tan notes, Thailand may well have the potential to become the center of a future transnational pan-Islamic struggle across the Malay Archipelago. Should this happen, "southern Thailand could then become Southeast Asia's Chechnya."[56] Despite grievances over land reform, governance, and corruption, the region has yet to become the second front of pan-Islamic instability. This may be true partly because of maritime geography and the Association of Southeast Asian Nations (ASEAN), which is focused on security cooperation among police and military forces. Yet as Rohan Gunaratna warns, eternal vigilance is vital because "there is no standard textbook for fighting al-Qaeda or its associated groups in Southeast Asia."[57]

STRATEGIC CULTURE AND AVENUES FOR ANALYSIS

The above surveys of irregular warfare in South and Southeast Asia indicate that much can be learned from studying the local dynamics and historical particularities of individual conflicts. At the microanalytical level, there can be little doubt that case histories and area studies continue to be indispensable tools for strategic studies. Accordingly, what are termed "mainstream strategic studies [as] an ethnocentric Anglo-American enterprise" clearly have much to gain from studying Asian culture and history.[58] It is less clear, however, whether analyzing classic works by Mao Zedong, Vo Nguyen Giap, and Abdul Nasution can provide novel theoretical insights on Asian irregular warfare at the macroanalytical level of strategic studies research.

While some observers such as Desmond Ball have speculated about the existence of Asia-Pacific strategic culture, there is little evidence to suggest that it has any relevance to the theory and practice of irregular warfare in Asia.[59] The concept of strategic culture might be interesting in examining China and India in light of Sun Tzu and Kautilya, but the search for a pan-Asian strategic culture will be futile. As Alastair Iain Johnston cautions, the distinctive should not be confused with the unique; done properly "analysis of strategic culture could help policymakers establish more accurate and empathetic understandings of how different actors perceive the game being played." But done badly, it "could reinforce stereotypes about the strategic dispositions of other states and close off policy alternatives deemed inappropriate."[60]

With the possible exception of China, war studies have not become a branch of systematic learning with empirical results as in the West.[61] The huge cultural and geographic diversity of Asia has resulted in a lack of regional strategic consciousness, which makes indigenous strategic cultures often parochial and particular. With respect to India, Kaushik Roy claims, "The Indian philosophy of war remains *terra incognita*" with little differentiation made between internal and external conflict.[62] By the same token, George Tanham attributes the absence of strategic theory to cultural inwardness, caste consciousness, and the fatalism of Hinduism.[63]

In sum, beyond analyses of individual nations, one should be skeptical about whether Asian strategic studies—particularly if couched in the argot of an Asian values school of international relations—can provide a touchstone for analyzing irregular warfare. As the late Gerald Segal put it, much of the Asian values school is simply a form of ethnic-chic conceived by intellectuals in China, Malaysia, and Singapore.[64] A similar conclusion is reached by Alan Dupont, who notes that "even though the [Asian] rhetoric of war and strategic discourse may be tinged by idiosyncratic cultural language . . . war fighting and dispute resolution . . . conforms to international, rather than specifically Asian, norms."[65] A recent, important work on irregular conflict, *The Routledge Handbook of Insurgency and Counterinsurgency*, remains fixed in the tradition of Western strategic studies. Actually, of thirty-one contributors, only three are from Asia, and specific cultural perspectives notwithstanding, each seems comfortable writing within a Western intellectual paradigm.[66]

Moreover, even the development of *jihadi* strategic studies, with its relevance to events in South and Southeast Asia, is largely a Western phenomenon. The term *jihadi strategic studies* was coined by Brynjar Lia and Thomas Hegghammer to isolate secular-rational ideas of al-Qaeda theoreticians such as Abu Mus'ab al-Suri and Abd al-Aziz al-Muqrin.[67] As Dima Adamsky suggests, the application of Western security studies has become necessary if we are to understand key aspects of globalized insurgency and terrorism. Al-Qaeda and Associated Movements may share a distinct strategic culture, global narrative, and common modes of warfare. These features may be reflected in theories of leaderless jihad that outline the "decentralization method [that] facilitates coexistence of strategic homogeneity and operational diversity" based on "system not organization" (*nizam la tanzim*).[68]

It remains the case in Asia that most contemporary scholars and practitioners of irregular conflict tend to draw on existing or emerging interdisciplinary models of Western strategic studies to explain irregular conflict. For example, recent studies of Indian and Sri Lankan insurgency have stressed the Western ideas on revolution and insurgency from Theda Scocpol and Bard O'Neill.[69] The reality suggests that improving comprehension of irregular warfare is not a quest for unique Asian strategic studies but rather the application of comparative analysis in the spirit of the Clausewitzian principle that good theory should always seek to educate the mind by revealing the distinctive elements of any military subject.[70]

With this Clausewitzian spirit in mind, employing four theoretical avenues of inquiry might best approach Asian irregular warfare. These avenues comprise, first, classical theory drawn from decolonization and the Cold War; second, neoclassical ideas to adapt classical methods to post–Cold War conditions; third, the concept of globalized insurgency and terrorism to discover new insights on transnational factors in irregular conflict; and fourth, stabilization theory for securing fragile states under threat from insurgency and civil insurrection.

Classical Avenue of Inquiry

This avenue concentrates on understanding the main characteristics of modern irregular warfare, including insurgency doctrines developed by Mao Zedong and Che Guevara. One of its main concerns is to identify the core ideas of modern COIN, such as the need for political primacy in strategy, civil-military unity, and the crucial difference between population-centric and enemy-centric modes of operation. Such ideas are discussed in works by such theorists as David Galula, Robert Thompson, and Frank Kitson, works that remain highly relevant in this century.[71] As noted previously, much subconventional warfare doctrine in 2006 relied on British COIN theory forged during the 1960s and 1970s.

In terms of understanding irregular warfare in Asia, the classical avenue, with its data on such campaigns as Malaya in the 1950s, has obvious applicability to ongoing ideological and class-based conflicts such as India's contemporary COIN effort against the Maoist Naxalites and the Philippine government's struggle against the Maoist New People's Army. In addition, the classical avenue provides a useful methodology to examine old ethnonationalist secessionist movements in South and Southeast Asia. For example, a separatist movement such as the Tamil Tigers, for all its tactical innovations, ultimately led into the kind of direct military confrontation that doomed earlier revolutionary insurgent movements such as the Greek Communists in the late 1940s. In short, the Tamil Tigers were destroyed by the application of an enemy-centric COIN campaign directed with firm political resolve. The conflict in Sri Lanka may be profitably studied from several classical perspectives, not least requirements for politicomilitary unity and large numbers of troops to fight guerrillas.

Similar perspectives might also apply to the Chinese COIN effort in Xinjiang. For instance, heavy-handed enemy-centric operations are being combined with society-centric methods such as large-scale Han Chinese immigration and economic development to suppress Uighur secessionists fighting for an East Turkestan state. The blend of military, political, and economic methods is redolent of a classical COIN approach.[72]

Neoclassical Avenue of Inquiry

The best approach to introducing the principles of insurgency and COIN in the post–Cold War era is the neoclassical avenue of inquiry that promotes organizational learning among practitioners.[73] Efforts like the Small Wars Operations Research Directorate (SWORD) Method usefully highlight the interaction of four factors in confronting insurgency: the character of host nations or client governments, strengths and weaknesses of insurgent organizations, capabilities of intervening powers or coalitions, and role of external sanctuaries in insurgencies.[74] Similarly, neoclassical analysis offers typologies of insurgent movements, including traditionalist, secessionist, egalitarian, commercialist, and criminal organizations, and their strategies. O'Neill's evaluation criteria for insurgency-counterinsurgency include the environment, popular support, organization, and external assistance, which are useful in understanding the evolution of irregular warfare since the 1990s. Finally, Mark Moyar's theory of insurgency and counterinsurgency as armed competition between superior elites that represents leader-centric warfare is a further contribution to comprehending irregular conflict.[75]

In terms of comprehending Asian irregular conflicts, these ideas are useful. For instance, the SWORD interaction theory, in which an internal-external center of gravity is established, is helpful in understanding the nexus between state and nonstate actors such as Pakistani military intelligence and the Lashkar-e-Toiba. Similarly, O'Neill's insurgent typology helps determine the identity of a bewildering number of different irregular movements that proliferate in such Southeast Asian nations as Indonesia, the Philippines, and Thailand. In the same vein, theories on elite-based leader-centric insurgency-counterinsurgency have obvious relevance in estimating the role of charismatic personality in movements such as al-Qaeda under Osama bin Laden, the Tamil Tigers under Prabarakhan, or Jemaah Islamiyah in Indonesia and Abu Bakar Bashir's important spiritual stewardship.

Global Avenue of Inquiry

The global avenue of inquiry considers irregular challenges as being transformed by the impact of globalization and information networks. Its focus is on the rise of a new form of transnational insurgency marked by the interaction between global-local matrices of irregular actors, including the relationship between al-Qaeda and the Quetta Shura Taliban along with the former's links to Abu Sayyaf in the Philippines and Jemaah Islamiyah in Indonesia.[76] The leading proponents of a jihadist global insurgency paradigm include John Mackinlay and David Kilcullen. The former advances the concept of an insurgent archipelago in the multicultural West, while the latter has developed a theory of the accidental guerrilla as a by-product of globalization and intervention within traditional societies to understand irregular threats.[77] Whereas classical counterinsurgency doctrines presupposed that the population from which the insurgents were likely to draw their recruits was concentrated in a single territory, newer transnational theories posit a diffuse and global battlefield.[78]

While accepting the value of studying classical paradigms of COIN and CT, Kilcullen believes that the irregular challenge is increasingly one of "a transfigured form of hybrid warfare that renders many of our traditional ideas irrelevant."[79] Thus, the global inquiry is clearly linked to *jihadi* strategic studies and conceptually important in understanding the use of proxy forces by Pakistan in Kashmir and India. Finally, a global perspective illuminates the ambitions of otherwise obscure Pakistan-based movements that include the Lashkar-e-Toiba, Lashkar-e-Janghvi, Sipa-e-Sohaba Pakistan, and Jaish-e-Mohammad.

Stabilization of Fragile States

The stabilization of the avenue of inquiry for fragile states was described by Francis Fukuyama, who argues, "State weakness is both a national and international issue of the first order."[80] From the perspective of strategic studies, a British doctrinal manual (*Stabilisation and Security*) and other works edited by David Richards and Greg Mills reveal the relationship between state weakness and military force.[81] The case is forcefully made in the British doctrinal manual that classical COIN and peace support operations are inadequate in managing twenty-first-century irregular challenges emerging from crises in ungoverned space. Instead, COIN efforts must be integrated into a wide-ranging approach to stabilization under globalized conditions, from military assistance to security and development. In this manner COIN becomes a refined tactical bubble to sustain population security, legitimate governance, and economic development.[82]

Insights from the nexus of stabilization and state building are useful in understanding how military forces might be employed as agents of modern conflict management, albeit in mainly a domestic rather than an expeditionary capacity. Some conceptual linkages in *Stabilisation and Security* might assist development of the Indian *Doctrine for Sub-Conventional Operations*, especially given the latter's strong emphasis on national unity, negotiated solutions, and political legitimacy. Moreover, stabilization and fragile-state theory remains pertinent to the many persistent internal conflicts in Southeast Asia being waged against centralization and the crisis of center-periphery governance in Pakistan, particularly given the tenuous authority of Islamabad over parts of its northwest border with Afghanistan.

• • •

South and Southeast Asia are challenged by irregular warfare. Although it is necessary to be sensitive to strategic cultures and distinctive national histories of irregular warfare, no indigenous strategic studies are available to analysts. Even with economic growth and advances in military modernization, remarkably little attention is dedicated in the region to the theoretical dimensions of insurgency or terrorism. Accordingly, the Western canon provides the most dynamic ideas concerning irregular warfare in South and Southeast Asia.

With respect to the future study of irregular conflict in Asia, the Western scholarly and policy communities are confronted with what might be termed a conceptual double blend. This double blend encompasses old and new operational-strategic modes as well as local and global political dynamics and occurs in a dual though interconnected interstate and substate security environment. Such a conceptual double blend is best comprehended through the perspective of Western strategic studies. In particular, the four avenues of classical, neoclassical, global, and stabilization and fragile-states theory suggest future ways to improve the understanding of irregular warfare in Asia in the twenty-first century. It remains a striking paradox that while South Asia is the center of gravity for dangerous insurgent and terrorist movements, most analyses originate in the West. From an Asian perspective, to quote Basil Liddell Hart, a situation prevails in which the challenges of irregular warfare remain "very long standing, yet [are] manifestly far from being understood."[83]

NOTES

1. John Baylis and James J. Wirtz, "Introduction," in *Strategy in the Contemporary World*, ed. John Baylis, James Wirtz, Colin S. Gray, and Eliot A. Cohen (Oxford: Oxford University Press, 2007), 1–14.

2. James D. Kiras, "Irregular Warfare: Terrorism and Insurgency," in Baylis, Wirtz, Gray, and Cohen, *Strategy in the Contemporary World*, 165.

3. Ibid.; Ariel Merari, "Terrorism as a Strategy of Insurgency," *Terrorism and Political Violence* 5 (Winter 1993): 213–51.

4. Douglas Blaufarb, *The Counterinsurgency Era: US Doctrine and Performance, 1950 to the Present* (New York: Free Press, 1977).

5. Raj Desai and Harry Eckstein, "Insurgency: The Transformation of Peasant Rebellion," *World Politics* 62 (July 1990): 441–65. For the background to the relationship between strategic theory and irregular warfare, see Charles Maechling, Jr., "Insurgency and Counterinsurgency: The Role of Strategic Theory," *Parameters* 14 (Autumn 1984): 32–41; Avi Kober, "Low-intensity Conflicts: Why the Gap Between Theory and Practise?" *Defense and Security Analysis* 18 (March 2002): 15–38; and James D. Kiras, "Irregular Warfare," in *Understanding Modern Warfare*, ed. David Jordan, James D. Kiras, David J. Lonsdale, Ian Speller, Christopher Tuck, and Dale C. Walton (Cambridge: Cambridge University Press, 2008), 224–68.

6. Desai and Eckstein, "Insurgency," 463–64.

7. John R. Galvin, "Uncomfortable Wars: Towards a New Paradigm," *Parameters* 26 (December 1986): 2–8.

8. Christopher Andrew, Richard J. Aldrich, and Wesley K. Wark, eds., *Secret Intelligence: A Reader* (London: Routledge, 2009), xv.

9. Thomas Rid and Thomas A. Keaney, "Understanding Counterinsurgency," in *Understanding Counterinsurgency: Doctrine, Operations and Challenges*, ed. Thomas Rid and Thomas A. Keaney (London: Routledge, 2010), 2.

10. United States Department of Defense, *Irregular Warfare: Countering Irregular Threats, Joint Operating Concept* (Washington, DC: Joint Chiefs of Staff, 2010), 5 (emphasis added).

11. United States Department of Defense, *Irregular Warfare: Joint Operating Concept* (Washington DC: Joint Chiefs of Staff, 2007), 5–6.

12. United States Department of Defense, "Irregular Warfare (IW)," Directive 3000.07, December 1, 2008, p. 2.

13. See Michael Evans, "Power and Paradox: Asian Geopolitics and Sino-American Relations in the Early Twenty-First Century," *Orbis: A Journal of World Affairs* 55 (Winter 2011), 85–113; Nick Beasley, *Building Asia's Security* (London: International Institute for Strategic Studies, 2009); and Amit Gupta, ed., *Strategic Stability in Asia* (Aldershot, UK: Ashgate, 2008).

14. See Saul Bernard Cohen, *Geopolitics: The Geography of International Relations*, 2nd ed. (Lanham, MD: Rowman and Littlefield, 2009), esp. chaps. 8–11.

15. See especially D. B. Shekatar, "India's Counterinsurgency Campaign in Nagaland," in *India and Counterinsurgency: Lessons Learned*, ed. Sumit Ganguly and David P. Fidler (New York: Routledge, 2009), 9–27; Vivek Chadha, "India's Counterinsurgency Campaign in Mizoram," in Ganguly and Fidler, *India and Counterinsurgency*, 28–44; and Walter C. Ladwig III, "Insights from the Northeast: Counterinsurgency in Nagaland and Mizoram," in Ganguly and Fidler, *India and Counterinsurgency*, 45–62; Ved Marwah, "India's Counterinsurgency Campaign in Punjab," in Ganguly and Fidler, *India and Counterinsurgency*, 89–106; and C. Christine Fair, "Lessons from India's Experience in the Punjab, 1978–93," in Ganguly and Fidler, *India and Counterinsurgency*, 107–26. See also Namrata Goswami, "Insurgencies in India," in *The Routledge Handbook of Insurgency and Counterinsurgency*, ed. Paul B. Rich and Isabelle Duyvesteyn (Abingdon, UK: Routledge, 2012), 208–17; and K. P. S. Gill, "Endgame in Punjab, 1988–93," in *Terror and Containment Perspectives of India's Internal Security*, ed. K. P. S. Gill and Ajai Sahni (New Delhi: Gyan, 2001), 23–84.

16. Fair, "Lessons from India's Experience," 110–11.

17. See Scott Gates and Kaushik Roy, introduction to *Unconventional Warfare in South Asia, 1947 to the Present*, ed. Scott Gates and Kaushik Roy (Farnham, UK: Ashgate, 2011), xvi–xxv; Jennifer L. Oetken, "Counterinsurgency Against the Naxalites in India," in Ganguly and Fidler, *India and Counterinsurgency*, 127–51; and Ajay K. Mehra, "Naxalism in India: Revolution or Terror?" *Terrorism and Political Violence* 12 (Summer 2000): 37–66.

18. See V. G. Patankar, "Insurgency, Proxy War, and Terrorism in Kashmir," in Ganguly and Fidler, *India and Counterinsurgency*, 65–78; Ashok Mehta, "India's Counterinsurgency Campaign in Sri Lanka," in Ganguly and Fidler, *India and Counterinsurgency*, 155–72; and Ryan Clark, *Laskkar-I-Taiba: The Fallacy of Subservient Proxies and the Future of Islamist Terrorism in India* (Carlisle Barracks, PA: Strategic Studies Institute, US Army War College, 2010).

19. Mehta, "India's Counterinsurgency Campaign," 164–65.

20. Patankar, "Insurgency," 65–66, 68–89; Sumit Ganguly, "Slow Learning: Lessons from India's Counterinsurgency Operation in Kashmir," in Ganguly and Fidler, *India and Counterinsurgency*, 79–88; Simon Jones, "India, Pakistan, and Counterin-

surgency Operations in Jammu and Kashmir," *Small Wars and Insurgencies* 19 (March 2008): 1–22.

21. Thomas Marks, "India: State Responses to Insurgency in Jammu and Kashmir—the Jammu Case," *Low Intensity Conflict and Law Enforcement* 12 (September 2004): 124.

22. Ibid., 126. Also see Clark, *Laskkar-I-Taiba*; and Adam Dolnik, "Fighting to the Death: Mumbai and the Future Fidayeen Threat," *RUSI Journal* 155 (April–May 2010): 60–68.

23. Namrata Goswami, "India's Counter-Insurgency Experience: The Trust and Nurture Strategy," *Small Wars and Insurgencies* 20 (March 2009): 66–86.

24. Sumit Ganguly and David P. Fidler, "Conclusion," in Ganguly and Fidler, *India and Counterinsurgency*, 225–26. See also David P. Fidler and Sumit Ganguly, "Counterinsurgency in India," in Rich and Duyvesteyn, *Routledge Handbook*, 301–11.

25. E. A. Vas, *Terrorism and Insurgency: The Challenge of Modernization* (Dehra Dun: Natraj, 1986), 219. See Dipankar Banerjee, "The Indian Army's Counterinsurgency Doctrine," in Ganguly and Fidler, *India and Counterinsurgency*, 189–206.

26. Ganguly, "Slow Learning," in Ganguly and Fidler, *India and Counterinsurgency*, 86.

27. Shekatar, "India's Counterinsurgency Campaign," 16–17.

28. R. Rajagopalan, "Force and Compromise: India's Counter-Insurgency Strategy," *South Asia: Journal of South Asian Studies* 30 (March 2007): 91. See also Vas, *Terrorism and Insurgency*, 227–28.

29. Patankar, "Insurgency," 76–77.

30. Shekatar, "India's Counterinsurgency Campaign," 18 Also see Banerjee, "The Indian Army's Counterinsurgency Doctrine," 189–206. On the role of the police, see Prem Mahadevan, "The Gill Doctrine: A Model for 21st Century Counter-Terrorism?" *Faultlines* 19 (April 2008): 1–14.

31. Republic of India, *Doctrine for Sub-Conventional Operations* (Simla: Headquarters, Army Training Command, 2006), i. See also David P. Fidler, "The Indian Doctrine for Sub-Conventional Operations: Reflections from a US Counterinsurgency Perspective," in Ganguly and Fidler, *India and Counterinsurgency*, 207–24.

32. Republic of India, *Doctrine for Sub-Conventional Operations*, 3, 15–16, 19, 27.

33. Ladwig, "Insights from the Northeast," 47. See also Goswami, "Insurgencies in India," 208–15.

34. See Shehzed H. Qazi, "Insurgent Movements in Pakistan," in Rich and Duyvesteyn, *Routledge Handbook*, 227–38; Bruce Riedel, *Deadly Embrace: Pakistan, America and the Future of the Global Jihad* (Washington DC: Brookings Institution Press, 2011); Rohan Gunaratna and Khuram Iqbal, *Pakistan: Terrorism Ground Zero* (London, Reaktion Books, 2011); Seth G. Jones and Christine Fair, *Counterinsurgency in Pakistan* (Santa Monica, CA: RAND, 2010); Ahmed Rashid, *Descent into*

Chaos: The US and the Disaster in Pakistan, Afghanistan, and Central Asia (London: Penguin Books, 2009); and Shuja Nawaz, *Crossed Swords: Pakistan, Its Army and the Wars Within* (Oxford: Oxford University Press, 2008).

35. Ayesha Siddiqa, "Pakistan's Counterterrorism Strategy: Separating Friends from Enemies," *Washington Quarterly* 34 (Winter 2011): 149–62.

36. Riedel, *Deadly Embrace*, 2.

37. Ibid., 33–47.

38. Akbar Ahmed, "With Obama at the World's 'Most Dangerous Place,'" *World Post*, March 28, 2009, http://www.huffingtonpost.com/akbar-ahmed/with-obamaat-the-worlds_b_180371.html.

39. Jones and Fair, *Counterinsurgency in Pakistan*, 24.

40. Shuja Nawaz, "Pakistan's Security and Civil-Military Nexus," in *The Afghanistan-Pakistan Theater*, ed. Daveed Gartenstein-Ross and Clifford D. May (Washington, DC: Foundation for Defense of Democracies Press, 2010), 18–27.

41. Riedel, *Deadly Embrace*, 85–96.

42. Syed Manzar Abbas Zaidi, "Pakistan's Anti-Taliban Counter-Insurgency," *RUSI Journal* 155 (February–March 2010): 10–19.

43. Owen Bennett Jones, "On the Verge: Pakistan's Insecurity," *RUSI Journal* 155 (February–March 2010): 4–8; Riedel, *Deadly Embrace*, 126–33.

44. Riedel, *Deadly Embrace*, 115.

45. Ashok Mehta, *Sri Lanka's Ethnic Conflict: How Eelam War IV Was Won* (New Delhi: Centre for Land Warfare Studies, 2010), 16–18; see also M. A. Thomas, "Sri Lanka: Political Uncertainty Under the Threat of Terrorism," in *Asian Security Handbook: Terrorism and the New Security Environment*, ed. William E. Carpenter and David G. Wiencek, 3rd ed. (New York: M. E. Sharpe, 2005), 273–82.

46. Mehta, "India's Counterinsurgency Campaign," 155–72; Chris Smith, "The LTTE: A National Liberation and Oppression Movement," in *Armed Militias of South Asia: Fundamentalists, Maoists and Separatists*, ed. Laurent Gayer and Christophe Jaffrelot (London: Hurst, 2009), 91–111.

47. Mehta, *Sri Lanka's Ethnic Conflict*, 1–4, 11.

48. Sinha Rajah Tammita-Delgoda, *Sri Lanka: The Last Phase in Eelam War IV from Chundikulam to Pudumattalan* (New Delhi: Centre for Land Warfare Studies, 2009), 23–24. See also David Lewis, "Counterinsurgency in Sri Lanka: A Successful Model?" in Rich and Duyvesteyn, *Routledge Handbook*, 312–34.

49. Rohan Gunaratna, "Terrorism in Southeast Asia: Threat and Response," in *A Handbook of Terrorism and Insurgency in Southeast Asia*, ed. Andrew T. H. Tan (Cheltenham, UK: Edward Elgar, 2007), 437–50. For useful overviews of insurgency and terrorism in Southeast Asia, see Gordon P. Means, *Political Islam in Southeast Asia* (Boulder, CO: Lynne Rienner, 2009), chaps. 9–11; Andrew T. H. Tan and Ken Booth, eds., *Non-Traditional Security Issues in Southeast Asia* (Singapore: Select Books, 2001);

Andrew T. H. Tan, *Security Perspectives of the Malay Archipelago: Security Linkages in the "Second Front" in the War on Terrorism* (Cheltenham, UK: Edward Elgar, 2004); and Greg Barton, *Jemaah Islamiyah Radical Islamism in Indonesia* (Singapore: Singapore University Press, 2005).

50. Milton Osborne, *Southeast Asia: An Introductory History*, 10th ed. (Sydney: Allen and Unwin, 2010).

51. Muthiah Alagappa, *The National Security of Developing States: Lessons from Thailand* (Dover, MA: Auburn House, 1987), 5; Means, *Political Islam in Southeast Asia*, chaps. 9–11; Andrew T. H. Tan, "Terrorism and Insurgency in Southeast Asia," in Tan, *Handbook of Terrorism and Insurgency*, 8–10.

52. Andrew T. H. Tan, "Old Terrorism in Southeast Asia: A Survey," in Tan, *Handbook of Terrorism and Insurgency*, 45–60.

53. Tan, "Terrorism and Insurgency," 5.

54. Ibid., 6–7; Tan, "Old Terrorism," 54; Paul A. Rodell, "Separatist Insurgency in the Southern Philippines," in Tan, *Handbook of Terrorism and Insurgency*, 225–47.

55. See Thitinan Pongsudhirak, "The Malay-Muslim Insurgency in Southern Thailand," in Tan, *Handbook of Terrorism and Insurgency*, 266–78; and Adam Dolnik, "Suicide Terrorism and Southeast Asia," in Tan, *Handbook of Terrorism and Insurgency*, 104–21. See also Rohan Gunaratna, Arabid Acharya, and Sabrina Chua, *Conflict and Terrorism in Southern Thailand* (Singapore: Marshall Cavendish, 2005).

56. Tan, "Old Terrorism," 58.

57. Gunaratna, "Terrorism in Southeast Asia," 447. For more on ASEAN security cooperation, see Tan, "Terrorism and Insurgency," 16–17.

58. Alan Macmillan, Ken Booth, and Russell Trood, "Strategic Culture," in *Strategic Cultures in the Asia-Pacific Region*, ed. Ken Booth and Russell Trood (London: Croom Helm, 1999), 17.

59. Desmond Ball, *Strategic Culture in the Asia-Pacific Region* (Canberra: Strategic and Defence Studies Centre, Australian National University, 1993), 21–22; Lawrence Sondhaus, *Strategic Culture and Ways of War* (London: Routledge, 2006), 89–119.

60. Alastair Iain Johnston, "Thinking About Strategic Culture," *International Security* 19 (Spring 1995): 63–64.

61. Kaushik Roy, "Just and Unjust War in Hindu Philosophy," *Journal of Military Ethics* 6 (2007): 232–45.

62. Ibid., 233.

63. George K. Tanham, "Indian Strategic Thought: An Interpretative Essay," in *Securing India: Strategic Thought and Practice: Essays by George K. Tanham with Commentaries*, ed. K. P. Bajpai and A. Mattoo (New Delhi: Manohar, 1996), 15–111; George K. Tanham, *The Ethics of War in Asian Civilization: A Comparative Perspective* (London: Routledge, 2006).

64. Gerald Segal, "What Is Asian About Asian Security?" in *Unresolved Futures: Comprehensive Security in the Asia-Pacific*, ed. James Rolfe, ed. (Wellington, New Zealand: Centre for Strategic Studies, 1995), 197.

65. Alan Dupont, "Is There an Asian Way?" *Survival* 38 (Summer 1996): 19.

66. See Rich and Duyvesteyn, *Routledge Handbook*, ix–xii. The Asian contributors are Indian and Singaporean.

67. See Brynjar Lia and Thomas Hegghammer, "Jihadi Strategic Studies: The Alleged Al Qaida Policy Study Preceding the Madrid Bombings," *Studies in Conflict and Terrorism* 27 (December 2004): 355–75; Steven Brooke, "Jihadist Strategic Debates Before 9/11," *Studies in Conflict and Terrorism* 31 (September 2008): 201–26; and Paul Cruickshank and Mohammad Hage Ali, "Abu Musab Al Suri: Architect of the New al-Qaeda," *Studies in Conflict and Terrorism* 30 (March 2007): 1–14. See also Brynjar Lia, *Architect of Global Jihad: The Life of Al-Qaida Strategist Abu Mus'ab al-Suri* (London: Hurst, 2007); and Norman Cigar, *Al-Qa'ida's Doctrine for Insurgency: 'Abd Al-'Aziz Al-Muqrin's* A Practical Course for Guerrilla War, trans. Norman Cigar (Washington DC: Potomac Books, 2009).

68. Dima Adamsky, "*Jihadi* Operational Art: The Coming Wave of *Jihadi* Strategic Studies," *Studies in Conflict and Terrorism* 33 (March 2010): 2, 9, 10–19.

69. The Indian studies include Durga Madhab (John) Mitra, *Understanding Indian Insurgencies: Implications for Counterinsurgency Operations in the Third World* (Carlisle Barracks, PA: Strategic Studies Institute, US Army War College, 2007); and S. P. Sinha, "Prabhakaran as Leader of the LTTE," *Journal of the United Services Institution of India* 131 (2001): 194–201. Mitra adopts ideas from Theda Scocpol, *Social Revolutions in the Modern World* (Cambridge: Cambridge University Press, 1994), while Sinha draws on Bard E. O'Neill, *Insurgency and Terrorism: From Revolution to Apocalypse* (Washington, DC: Potomac Books, 2005).

70. Carl von Clausewitz, *On War*, ed. Michael Howard and Peter Paret (Princeton, NJ: Princeton University Press, 1976), 141.

71. David Galula, *Counterinsurgency Warfare: Theory and Practice* (Westport, CT: Praeger, 1964); Robert G. K. Thompson, *Defeating Communist Insurgency: Experiences from Malaya and Vietnam* (London: Chatto and Windus, 1966); Frank Kitson, *Low Intensity Operations: Subversion, Insurgency, Peacekeeping* (London: Faber and Faber, 1971).

72. Martin I. Wayne, *China's War on Terrorism: Counterinsurgency, Politics and Internal Security* (London: Routledge, 2008); Martin I. Wayne, "China's Society-Centric Counterterrorism Approach in Xinjiang," in Rich and Duyvesteyn, *Routledge Handbook*, 335–46.

73. See Max G. Manwaring and John T. Fishel, "Insurgency and Counter-Insurgency: Towards a New Analytical Approach," *Small Wars and Insurgencies* 3 (Winter 1992): 272–93; O'Neill, *Insurgency and Terrorism*; John A. Nagl, *Counterinsurgency Lessons*

from Malaya and Vietnam: Learning to Eat Soup with a Knife (Westport, CT: Praeger, 2002); Robert M. Cassidy, *Counterinsurgency and the Global War on Terror: Military Culture and Irregular War* (Westport, CT: Praeger, 2006); and Mark Moyar, *A Question of Command: Counterinsurgency from the Civil War to Iraq* (Cambridge, MA: Yale University Press, 2009).

74. Manwaring and Fishel, '"Insurgency and Counter-Insurgency."

75. O'Neill, *Insurgency and Terrorism*, 199–203; Moyar, *A Question of Command*, 1–15.

76. Jalil Roshandal and Sharon Chadha, *Jihad and International Security* (New York: Palgrave Macmillan, 2006).

77. John Mackinlay, *The Insurgent Archipelago: From Mao to Bin Laden* (New York: Columbia University Press, 2009), chaps. 5–8; David Kilcullen, *The Accidental Guerrilla: Fighting Small Wars in the Midst of a Big One* (Oxford: Oxford University Press, 2009), chaps. 1 and 5.

78. Christopher Coker, *Ethics and War in the 21st Century* (London: Routledge, 2008), 101–102.

79. Kilcullen, *The Accidental Guerrilla*, xvii.

80. Francis Fukuyama, *State-Building: Governance and World Order in the 21st Century* (Ithaca, NY: Cornell University Press, 2004), x–xi; Francis Fukuyama, ed., *Nation-Building: Beyond Afghanistan and Iraq* (Baltimore, MD: Johns Hopkins University Press, 2006).

81. United Kingdom Ministry of Defence, *Stabilisation and Security: The Military Response*, Joint Doctrine Publication 3-40 (Shrivenham, UK: Development, Concepts and Doctrine Centre, 2009); David Richards and Greg Mills, *Victory Among People: Lessons from Countering Insurgency and Stabilising Fragile States* (London: Royal United Services Institute, 2011).

82. United Kingdom Ministry of Defence, *Stabilisation and Security*, 2-15–2-30.

83. B. H. Liddell Hart, *Strategy*, 2nd ed. (New York: Meridian, 1991), xv.

CONCLUSION

Toward a Research Agenda

Thomas G. Mahnken, Dan Blumenthal, and Michael Mazza

THIS VOLUME IS BASED on two premises. The first is that the increasing strategic weight of Asia demands greater attention from scholars and policy makers alike. The second is that a full understanding of the rising importance of Asia requires that it be viewed under multiple lenses: those of geography, politics, economics, history, and culture. Many of the ideas included in the strategic studies canon are similarly useful, such as deterrence and arms race. However, the volume marks only an initial effort to promote a line of inquiry intended to explicitly link the fields of strategic studies and Asian affairs. Each chapter raises almost as many questions as it attempts to answer, which can serve as points of departure for both academics and practitioners.

GEOGRAPHY AND STRATEGY

The relationship between geography and strategy is enduring. Conflicts between continental and maritime powers have occurred throughout history. For example, competition between Athens and Sparta, Rome and Carthage, Britain and France (and later, Germany), and the United States and Soviet Union was influenced by geography. In the chapter "The Cyclical Nature of Chinese Sea Power," Bruce Elleman argues that China has shifted from being a continental to a maritime power and back again as circumstances dictated. In other words, geography has not been destiny, at least in the case of China. This claim made by Elleman will doubtless stimulate further debate since it bears immediately on understanding the contemporary determination by China to become a sea power and, potentially, the limits of such efforts.

Additional studies should consider the role of geography in Asian history. For example, the visit by Commodore Matthew Perry to Japan is generally regarded as the point that separates the isolated past of the nation from its eventual outward-looking history. Nevertheless, many factors contributed to this development. How did the culture and geography of Japan manage to keep the outside world at bay for so long? How did interaction by Japan enable it to venture abroad during the last part of the nineteenth and first half of the twentieth centuries? Has Japan been amenable to maintaining its postwar neutering in more recent times because it is an island nation?

Natural and imposing land barriers in China—the Gobi Desert and Tibetan plateau—did not prevent it from expanding in earlier times. Though the Chinese were capable of projecting power over great distances at sea, their maritime expansion tended to be *relatively* short-lived. Interestingly, far-flung vestiges of the British and French empires have long outsurvived their other possessions while the Chinese have held on to none of their own. Why?

In the chapter "Chinese Maritime Geography," Toshi Yoshihara asserts that "to Beijing, the *first island chain*—the long ring of islands lying just off the eastern end of Eurasia—hems in the Chinese mainland. This geographic view is integral to Chinese strategic thinking on sea power." But how does the United States, and especially Japan, look at this island chain, of which Japan is part? Where China perceives threat, does Japan see opportunity? Does Japan regard the islands to the south and west of Kyushu as important or even defining in East Asian maritime geography? Additionally, have the respective interpretations of China and Japan of their shared maritime geography shifted over time? Finally, are there reasons to expect that these Chinese and Japanese views will change in the future?

STRATEGIC CULTURE

Culture serves as a powerful lens through which to fathom strategic behavior. Colin Gray in his chapter "Strategy and Culture" suggests that the concept of strategic culture is flawed but useful. And he concedes that while its explanatory powers are limited, the quality of insight it offers is beyond serious challenge. Moreover, he notes something obvious but sadly not widely known—that strategy "involves life and death writ large; ideas matter because of their potential impact on people. Thinking strategically is a practical endeavor, which means that attitudes, beliefs, and values with cultural import should not be gauged in terms of logic alone."

Examinations of strategic culture begin with a definition. For Gray it is "the assumptions, beliefs, attitudes, habits of mind, and preferred modes of behavior, customary behavior even, bearing on the use of force by a security community." This is a succinct version of a definition proposed by Alastair Iain Johnston that envisions strategic culture as an integrated

> system of symbols (e.g., argumentation structures, languages, analogies, metaphors) which acts to establish pervasive and long-lasting strategic preferences by formulating concepts of the role and efficacy of military force in interstate political affairs, and by clothing these conceptions with such an aura of factuality that the strategic preferences seem uniquely realistic and efficacious.[1]

Johnston's definition is instructive in that it places the assumptions, beliefs, attitudes, habits of mind, and modes of behavior as given by Gray in context. Nonetheless, identifying which aspects of culture affect strategy making and which do not, and why, presents a significant challenge for anyone who argues for the existence of distinct strategic cultures.

China

As Andrew Wilson suggests in "The Chinese Way of War," the dominant view of Chinese strategic culture has little basis in Chinese military history. Indeed, if it did, the military history of China might look quite different. As outsiders attempt to predict China's strategic choices, they must determine whether it is useful to look back at history—and if so, what aspects of that history—or interpret the modern Chinese record.

This paradox suggests two research approaches. First, research should determine where in modern Chinese history the narrative of five myths as described by Wilson emerged. Then it should identify those instances when China used force and compare the explanatory power of strategic tradition with a more recent conception of strategic culture. Next, this research must determine whether these contradictions are universal. Are some cultures better at understanding their strategic traditions than others? Is the interaction among nations that accurately recognize strategic culture different from those that do not? Are nations that misconstrue their own strategic cultures more likely to misunderstand the strategic cultures of others? If true, what are the implications for strategy making and the success of the strategy? Finally, future research should examine the organizational culture of the People's Liberation

Army and its components. China, for instance, traditionally prefers secrecy and stratagem. The PLA notion of deterrence also differs markedly from Western strategy.[2] Such questions deserve greater scholarly attention.

Japan

S. C. M. Paine raises interesting questions about the relationship between culture and strategy in her chapter "The Japanese Way of War." She observes that during World War II "there were cultural explanations for the Japanese neglecting grand strategy, refusing to cut losses and end the conflict, failing to coordinate operations, and behaving with such ferocity." She states that *bushido*, the Samurai code, deals with death; features honor, suicide, loyalty, and willpower; and denigrates the utility of strategy. In wartime, *bushido* led to neglect of logistics and sea-lanes, rivalries between and within the army and navy, and failure to terminate the war when it became clear that victory was no longer a possibility.

Perhaps the most important observation offered by Paine is that the Japanese way of war provides "a cautionary tale for those who make predictions based on an assumed monopoly on rationality that appears in terms of the universality of their own values." Western thinkers—economists, politicians, soldiers—have long made assumptions about rational actors and how people behave. Strategists in particular must reexamine how they perceive rationality. Rational actors in Washington, Tokyo, Beijing, and elsewhere might make different choices under similar circumstances and pursue different ends. The ability to predict decisions would contribute to enhanced strategy making. If the United States had understood at the outset of the Pacific War that death would be preferable to life after defeat for many Japanese, would that have altered the American conduct of military operations?

Paine delves into the connection between culture and the outcome of the war. She suggests that culture provided one reason for the ultimate defeat of Japan. However, at the time, modern Japanese history was replete with military successes. If culture can be blamed only in part for the failures by Japan in World War II, can it be credited for its victories in the Russo-Japanese and Sino-Japanese Wars only a few decades earlier? And is culture responsible for the Japanese success in seizing Manchuria from China in 1931 and establishing Manchukuo the following year? Was the Japanese strategy during the first half of the twentieth century effective because of or in spite of Japanese strategic culture?

Other Powers

Most contributors to this volume focused on the great powers in the Asia-Pacific region: China, Japan, India, and the United States. Yet other smaller, even rising powers are likely to exercise considerable influence in the future, especially Australia, South Korea, Indonesia, Vietnam, and the Philippines. For the United States, those nations include three allies, one former enemy, and one state whose relationship has waxed and waned over recent decades. Their strategic cultures are even less understood than those of China and Japan, with the possible exception of Australia. Some possess advanced military capabilities, most are anxious about China, some have disputes with one another, and all seek closer partnerships with the United States. Appreciating how each nation perceives threats and makes strategy is important for the future.

Australia

Although it shares many traditions with Great Britain, Australia in some ways may have more in common with the United States. Australians have fought alongside Americans in major conflicts since World War I—World War II, Korea and Vietnam, and Afghanistan and Iraq. Moreover, Australia has acted much closer to home, intervening in East Timor and the Solomon Islands and engaging in counterterrorism efforts in Indonesia, which is evocative of US involvement in Central America and the Caribbean. Strangely, the activist strategic impulse of Canberra seems at odds with its ostensible security. Although Australia remains concerned over Indonesia and Papua New Guinea, it is the only continental nation far removed from global hot spots.

How is Australian strategic culture affected by its historical links to Great Britain and its more recent security relationship with the United States? What role has Australian geography played in its security outlook and the development of its strategic culture?[3] Is the continental power–maritime power dichotomy relevant? Michael Evans, for instance, has noted elsewhere that Australia lacks a maritime identity.[4] What in the record of Australian strategy making can help in predicting how Australia will approach China in the future?

South Korea

Since it was established in 1948, the Republic of Korea (ROK) has faced one dominant security concern: the Democratic People's Republic of Korea (North Korea). Conditions on the Korean Peninsula have changed with the demise of the Soviet Union, the opening of China, and the vanished threat of Japan, but Pyongyang commands the attention of the national security establishment in Seoul. The Korean Peninsula is situated among three behemoths—China, Japan, and Russia. Korea was long a tributary state of China. Japan's desired expansionism on the Peninsula, which was realized before the 1900s, dates back many centuries. And in more recent times, Russia has eyed Korea for its geographic potential to provide warm-water ports.

Is Korean strategic culture markedly different from that of China, which shares Confucian values and was once the suzerain of the country? Did either the birth of the Democratic People's Republic of Korea in 1948 or the Korean War serve as a turning point in the South Korean ways of strategy making? Does the history of Korea as a Chinese tributary state reveal lessons on how it might deal with the United States, the dominant member of the US-ROK security alliance? Does history proffer lessons on how South Korea will strategize vis-à-vis China in the future? Has the alliance with Washington changed the way that Seoul makes strategy?

Indonesia

Like the Philippines, Indonesia has a comparatively brief history as a unitary, independent nation. And like the Philippines, its security interests have focused on internal stability. Externally, however, Indonesia has been largely preoccupied by rivalry with Singapore and Malaysia, although lately its relations with both parties have become cooperative rather than competitive. More recently, Indonesia has expressed concern over potential regional instability caused by the rise of China and tension over territorial disputes. Nonetheless, its strategy of *dynamic equilibrium* for handling such concerns appears to be focused on the ends, to the exclusion of ways and means.[5] Does Indonesian history reveal a discernable way of making strategy? Does its limited experience in interstate conflict hamper its ability to deal with growing Asian tensions? If Jakarta succeeds in ensuring internal security, how will its approach to the outside world shift?

Vietnam

The history of the Vietnamese people is replete with narratives on resistance to intervention by foreign powers: China, France, Japan, America, and China once again. Today, problems in its neighborhood worry Hanoi, especially Chinese military modernization, which in some ways is old hat. Recent overtures to the United States are consistent with modern Vietnamese history:

seeking support from a strong, ideally distant power to counter immediate threats. Vietnam is similar to China in many ways, but its military history is different. However, do these nations have similar strategic cultures? While Vietnam is part of mainland Southeast Asia, its fortunes are increasingly tied to maritime or insular Southeast Asia. How has that dual identity come to shape Vietnamese strategy making? Even though naval forces play second fiddle to the ground forces within the Vietnam People's Army, the situation appears to be shifting as Hanoi confronts Beijing over rights in the South China Sea. How will this focus on maritime power affect future Vietnamese strategy?

The Philippines

Of all the allies of the United States within the region—Australia, Japan, the Philippines, South Korea, and Thailand—the Philippines has the shortest history as a unified political unit. But significant cultural differences exist across the islands, and the country's past has been marked by diverse influences, both cultural (Chinese, Spanish, and American) and religious (Islam and Catholicism). Until recently, the Philippines has been preoccupied with internal challenges and its legitimacy as a unified nation. While disputes with neighbors in the South China Sea are not new, they only intermittently dominate the national security agenda.

Do the Philippine Islands have a unique strategic culture? If so, does it have origins in the pre-Conquista period, before the forceful unification of numerous kingdoms and sultanates? Have external historical influences molded its strategic culture? Has a long-insular focus ill equipped Manila to make strategy for an increasingly unstable environment?

IRREGULAR WARFARE

Michael Evans closely examines what has become accepted as counterinsurgency in South and Southeast Asia in his chapter "Irregular Warfare in Asia." He writes that Asian strategies for counterinsurgency depend on Western scholarship for their intellectual bases. "Remarkably little attention is dedicated in the region to the theoretical dimensions of insurgency or terrorism. Accordingly, the Western canon provides the most dynamic ideas concerning irregular warfare in South and Southeast Asia." Further research should investigate this apparently regional disinterest in the systematic analysis of irregular warfare. Evans implies that it is important to make an investment in such research:

> South Asia is the center of gravity for dangerous insurgent and terrorist movements, [and] most analyses originate in the West. From an Asian perspective, to quote Basil Liddell Hart, a situation prevails in which the challenges of irregular warfare remain "very long standing, yet [are] manifestly far from being understood."[6]

The insurgents themselves might be useful as a starting point. Asia has had a long and rich history of insurgencies, which are not limited to recent trouble spots. While the thoughts of Mao Zedong on people's war have been well rehearsed, less attention has been given to Kuomintang insurgencies at the end of the Chinese Civil War and to the two-decade-long insurgency in Tibet. How did Mao make the transition from mounting insurgencies to squashing them? Did his strategy stem from personal experience, or did he rely on foreign lessons to subdue uprisings? Elsewhere, before the Korean War, Jeju Island played host to a revolt, which South Korean forces brutally suppressed. During the war itself, North Korean soldiers and local civilians in the South engaged in guerrilla warfare. Indigenous groups resisted Japanese rule on Taiwan for decades. Analyzing how such conflicts were handled before the counterinsurgency era may illuminate any surviving Asian traditions in combating insurgencies and waging irregular war.

Western militaries have been heavily involved in recent insurgencies in Asia. The United States has been engaged in counterinsurgency warfare in Afghanistan for a decade, including along the border with Pakistan. It also has been supporting counterinsurgency efforts on Mindanao during the same time with military assistance and training. Moreover, Australia has been extensively supporting comparable operations in Indonesia.

Future research must examine how the Indonesian experiences, in which Western nations are not in the lead, are shaping counterinsurgency and counterterrorism theories and praxis. By adopting methods developed in the West, does Indonesia or the Philippines implement them in indigenous ways? If so, what can be learned by this adaptation of Western ways?

• • •

As Aaron Friedberg notes in his introduction, some of the greatest challenges the United States will confront in the coming decades are likely to originate in the Asia-Pacific region. China and India are rising powers, militant Islamists still thrive across Pakistan, and nuclear proliferation remains unrestrained.

To meet these challenges, American strategists must endeavor to learn more about this region much as specialists in Asian affairs must come to the study of the enduring principles of strategy.

NOTES

1. Alastair Iain Johnston, "Thinking About Strategic Culture," *International Security* 19 (Spring 1995): 46. Johnston acknowledges paraphrasing a definition of religion as a cultural system by Clifford Geertz in *The Interpretation of Cultures* (New York: Basic Books, 1973), 90.

2. Dean Cheng, "Chinese Views on Deterrence," *Joint Force Quarterly* 60 (2011): 92–94.

3. Michael Evans, *The Tyranny of Dissonance: Australia's Strategic Culture and Way of War, 1901–2005* (Canberra: Land Warfare Studies Center, 2005).

4. Michael Evans, "The Challenge of Australian Maritime Identity," *Quadrant*, November 2013, pp. 22–30.

5. See Marty Natalegawa, minister of Foreign Affairs, Republic of Indonesia, interview by the Council on Foreign Relations, broadcast September 20, 2010: "Our interest is what we call a dynamic equilibrium for our region, not quite a classic balance-of-power situation where not one nation is preponderant in our region, but in a more holistic and a more hopefully positive sense, in the sense that we don't wish to see our region dominated by one nation."

6. B. H. Liddell Hart, *Strategy*, 2nd ed. (New York: Meridian, 1991), xv.

ABOUT THE CONTRIBUTORS

Michael R. Auslin directs Japan studies at the American Enterprise Institute. Previously he was an associate professor and senior research fellow in the MacMillan Center for International and Area Studies at Yale University. His books include *Pacific Cosmopolitans: A Cultural History of US-Japan Relations* and *Security in the Indo-Pacific Commons: Toward a Regional Strategy*. Dr. Auslin has been named a Young Global Leader by the World Economic Forum, a Marshall Memorial Fellow by the German Marshall Fund, a Fulbright Scholar, and a Japan Foundation Scholar.

Richard A. Bitzinger is a senior fellow in the S. Rajaratnam School of International Studies at Nanyang Technological University. He has published articles in *International Security*, *Journal of Strategic Studies*, *Orbis*, and *Survival* and is the editor of *The Modern Arms Industry: Political, Economic, and Technological Issues*. Mr. Bitzinger taught at the Asia-Pacific Center for Security Studies and worked for RAND and the Center for Strategic and Budgetary Assessments. In addition, he was a senior fellow at the Atlantic Council of the United States.

Dan Blumenthal directs Asian studies at the American Enterprise Institute and serves on the US-China Economic and Security Review Commission and the Academic Advisory Board for the congressional US-China Working Group. Previously he was the senior director for China, Taiwan, and Mongolia in the Office of the Secretary of Defense. Mr. Blumenthal is the author with Phillip Swagel of *An Awkward Embrace: The United States and China in the 21st Century* and has contributed to the *Wall Street Journal*, the *Weekly Standard*, and *National Review*.

Michael S. Chase is a senior political scientist at RAND. He also teaches at the Paul H. Nitze School at the Johns Hopkins University. Previously he was an associate research professor in the Warfare Analysis and Research Department and director of the Mahan scholars at the Naval War College. Dr. Chase wrote *Taiwan's Security Policy: External Threats and Domestic Politics* and published a number of book chapters on China and Asian security issues. He has published articles in *Asia Policy, Joint Force Quarterly, Journal of Strategic Studies, Proceedings,* and *Survival.*

Bruce A. Elleman is a research professor with the Maritime History Department in the Center for Naval Warfare Studies at the Naval War College. His publications include *Naval Power and Expeditionary Wars: Peripheral Campaigns and New Theatres of Naval Warfare*; *Nineteen Gun Salute: Case Studies of Operational, Strategic, and Diplomatic Naval Leadership During the 20th and Early 21st Centuries*; *Piracy and Maritime Crime: Historical and Modern Case Studies*; *Manchurian Railways and the Opening of China: An International History*; and *Modern China: Continuity and Change, 1644 to the Present.*

Michael Evans is the General Sir Francis Hassett Chair of Military Studies at the Australian Defence College and a professor at Deakin University. Formerly, he headed the Land Warfare Studies Centre at the Royal Military College. Earlier, he served with the Rhodesian Army and the Zimbabwe National Army, integrating guerrilla cadres into a professional force. Dr. Evans was responsible for counterinsurgency education in the Australian Defence Force and wrote the Australian Army manual Land Warfare Doctrine 3-0-1, *Counterinsurgency.*

Aaron L. Friedburg is a professor of politics and international affairs at Princeton University. Previously he directed research at the Woodrow Wilson School and the Center of International Studies at Princeton. Dr. Friedberg also has served as the Deputy Assistant for National Security Affairs in the Office of the Vice President. He has held fellowships at the Center of International Affairs at Harvard and the Woodrow Wilson International Center for Scholars. His latest book is *A Contest for Supremacy: China, America, and the Struggle for Mastery in Asia.*

Colin S. Gray is a professor of international politics and strategic studies and also the director of the Centre for Strategic Studies at the University of Reading. He has held posts at Hull, Lancaster, and York Universities. Dr. Gray was affiliated earlier with the International Institute of Strategic Studies, Hudson Institute, and National Institute for Public Policy. Among his recent books are *Strategy and History: Essays on Theory and Practice* and *War, Peace, and International Relations: An Introduction to Strategic History.*

James R. Holmes is an associate professor of strategy at the Naval War College. He previously taught at the University of Georgia and also worked as a researcher with the Institute for Foreign Policy Analysis. As a surface warfare officer, he served aboard USS *Wisconsin* and was assigned to both the Surface Warfare Officers School Command and the College of Distance Education of the Naval War

College. His latest book is *Red Star over the Pacific: China's Rise and the Challenge to US Maritime Strategy.*

Timothy D. Hoyt is a professor of strategy and policy at the Naval War College and has taught in the School of Foreign Service at Georgetown University. He has worked for the Department of the Army and the Congressional Research Service. Dr. Hoyt cochairs the Indian Ocean Regional Studies Group at Newport, Rhode Island, and participated in track II discussions with India and Pakistan. He is the author of *Military Industries and Regional Power* and is writing two books, on US military strategy and the Irish Republican Army.

Roy D. Kamphausen is senior associate for political and security affairs at the National Bureau of Asian Research. A former US Army foreign area officer, he served with the Defense Attaché Office in Beijing. He contributed the chapters "Military Modernization in Taiwan" to *Strategic Asia 2005–06: Military Modernization in an Era of Uncertainty* and "PLA Power Projection: Current Realities and Emerging Trends" to *Assessing the Threat: The Chinese Military and Taiwan's Security*, and he edited *The PLA at Home and Abroad: Assessing the Operational Capabilities of China's Military.*

Bradford A. Lee holds the Philip A. Crowl Chair of Comparative Strategy at the Naval War College and has taught at Harvard University. He coedited a book of essays, *Strategic Logic and Political Rationality*, and contributed chapters titled "The Cold War as a Coalition Struggle" to *Naval Coalition Warfare: From the Napoleonic War to Operation Iraqi Freedom* and "Teaching Strategy: A Scenic View from Newport" to *Teaching Strategy: Challenge and Response.* Currently he is finishing a book, *On Winning Wars.*

Thomas G. Mahnken holds the Jerome E. Levy Chair of Economic Geography and National Security at the Naval War College and is a senior research professor at the Paul H. Nitze School of Advanced International Studies at the Johns Hopkins University. Formerly he was the Deputy Assistant Secretary of Defense for Policy Planning. Dr. Mahnken is the author of *Technology and the American Way of War Since 1945* and the editor of *Competitive Strategies for the 21st Century: Theory, History, and Practice.* In addition, he edits the *Journal of Strategic Studies.*

Michael Mazza manages the executive program on national security policy and strategy at the American Enterprise Institute, where he focuses on US defense policy in the Asia-Pacific region and on Chinese military modernization, Korean security affairs, and cross-strait relations. Mr. Mazza also contributes to the Center for Defense Studies blog. He was named a Foreign Policy Initiative Future Leader (2010–2011) and earlier worked as research analyst at Riskline Limited, Hicks and Associates, and the Southeast Asia Program at Cornell University.

S. C. M. Paine is a professor of strategy and policy at the Naval War College. Her publications include *The Sino-Japanese War of 1894–1894: Power, Perceptions, and Primacy* and *Imperial Rivals: China, Russia, and Their Disputed Frontier, 1858–1924*. Dr. Paine also has written and edited a number of books, including *Modern China: Continuity and Change, 1644 to the Present*; *Naval Coalition Warfare: From the Napoleonic War to Operation Iraqi Freedom*; and *Naval Blockades and Seapower: Strategies and Counter-Strategies, 1805–2005*.

Andrew R. Wilson is a professor of strategy and policy at the Naval War College. Previously he taught at Wellesley College and Harvard University. He has published numerous articles on Chinese military history, Chinese sea power, and Sun Tzu. In addition, he has written books on Chinese communities in the Philippines and the Caribbean. Recently he edited *China's Future Nuclear Submarine Force* and a multivolume history of the Sino-Japanese War (1937–1945), and he is completing a new translation of *The Art of War* by Sun Tzu.

Toshi Yoshihara holds the John A. van Beuren Chair of Asia-Pacific Studies and is affiliated with the China Maritime Studies Institute at the Naval War College. Earlier he taught at the Air War College and worked for RAND and the American Enterprise Institute. Dr. Yoshihara is the author of *Red Star over the Pacific: China's Rise and the Challenge to US Maritime Strategy* and *Indian Naval Strategy in the Twenty-first Century*, and he is the editor of *Asia Looks Seaward: Power and Maritime Strategy*.

INDEX

Page numbers in italics indicate material in figures.